ROYAL SOCIETY OF CHEMISTRY

# Metal–Ligand Bonding

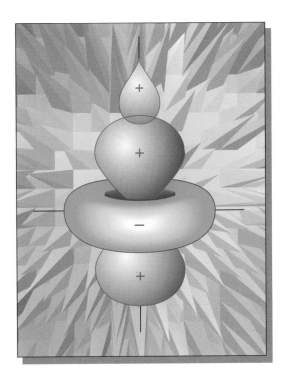

Rob Janes
*and*
Elaine Moore

This publication forms part of the Open University course S343 Inorganic Chemistry. Details of this and other Open University courses can be obtained from the Course Information and Advice Centre, PO Box 724, The Open University, Milton Keynes MK7 6ZS, United Kingdom: tel. +44 (0)1908 653231, e-mail general-enquiries@open.ac.uk.

Alternatively, you may visit the Open University website at http://www.open.ac.uk, where you can learn more about the wide range of courses and packs offered at all levels by The Open University.

The Open University
Walton Hall, Milton Keynes
MK7 6AA

First published 2004.

Edited, designed and typeset by The Open University.

Published by the Royal Society of Chemistry, Thomas Graham House, Science Park, Milton Road, Cambridge CB4 0WF, UK.

Registered Charity Number 207890.

Printed and bound in the United Kingdom by Bath Press, Glasgow .

ISBN 0 85404 979 7

Orders and Enquiries should be sent to:

Sales and Customer Care Department, Royal Society of Chemistry, Thomas Graham House, Science Park, Milton Road, Cambridge, CB4 0WF, UK

Tel: +44 (0)1223 432360; Fax +44 (0)1223 426017; e-mail: sales@rsc.org

1.1

s343block 2 i1.1

# PREFACE

This book aims to provide an accessible description of the theory of transition metal–ligand bonding, written in a detailed, yet non-mathematical manner. The way bonding models can be used to rationalise many of the chemical and physical properties of complexes is emphasised throughout. The text begins with a brief consideration of the electronic configuration of d electrons on metal ions and the anatomy of a complex, leading to a discussion of the delightfully simple yet extremely powerful crystal-field model. Using this model, we then describe the use of magnetic measurements to distinguish complexes with different electronic configurations and geometries. With basic symmetry concepts as a foundation, this is followed by a treatment of molecular orbital theory applied to transition-metal complexes, using a pictorial approach. Emphasis is placed on the relationship between crystal field and molecular orbital theories throughout. Both d–d and charge-transfer spectra are used to link theory to observation. Even though this text is centred on theoretical models, we have endeavoured to emphasise the practical relevance of the material by the inclusion of relevant experimental data and observations from everyday life.

Full colour energy-level diagrams and orbitals are used throughout. Included in the text are learning outcomes for each section, embedded questions (with answers), and revision exercise questions emphasising connections between different areas of the text. A basic knowledge of atomic and molecular orbitals as applied to main group elements is assumed.

Many people helped with the production of this book. We should like to thank Margaret Careford for word processing, Pam Owen for turning our rough sketches into handsome illustrations, Mike Levers for his high-quality photographs, Ian Nuttall for his thorough editing, Jane Sheppard for cover design and layout, our colleagues Dr Charlie Harding and Yvonne Ashmore, Dr Chris Jones of BNFL for helpful comments, and the RSC for their faith in agreeing to co-publish this text.

*Rob Janes*
*Elaine Moore*

# CONTENTS

# INTRODUCTION

The attribute of transition-metal ions on which this book focuses is their possession of partially occupied **d orbitals**. Across the fourth row of the Periodic Table, an electron enters the 4s sub-shell at potassium, and a second fills it at calcium. Then, from scandium to zinc, the 3d sub-shell is progressively filled. For the neutral atoms, the energies of the 3d and 4s orbitals are very close, and it is the *exchange energy stabilisation* [†] associated with half-filled and filled shells that gives rise to configuration irregularities at chromium and copper, respectively. This is shown in Table 1.1, where [Ar] represents the argon core electrons.

When transition-metal atoms form cations, the 4s electrons are lost first. On ionisation, the 3d orbitals are significantly more stabilised (that is, drop to lower energy) than the 4s would be. This stems from the fact that the 3d electrons are not shielded from the nucleus as well as the 4s electrons. Therefore, the +2 and +3 ions have electronic configurations of $[Ar]3d^n$ (or $1s^22s^22p^63s^23p^63d^n$). The electronic configurations of the +2 and +3 ions, which we shall refer to frequently, are shown in Table 1.2.

**Table 1.1** Electronic configurations of the free atoms of the first transition series and zinc

| Element | Configuration |
|---------|---------------|
| Sc | $[Ar]3d^14s^2$ |
| Ti | $[Ar]3d^24s^2$ |
| V | $[Ar]3d^34s^2$ |
| Cr | $[Ar]3d^54s^1$ |
| Mn | $[Ar]3d^54s^2$ |
| Fe | $[Ar]3d^64s^2$ |
| Co | $[Ar]3d^74s^2$ |
| Ni | $[Ar]3d^84s^2$ |
| Cu | $[Ar]3d^{10}4s^1$ |
| Zn | $[Ar]3d^{10}4s^2$ |

**Table 1.2** Electronic configurations of the dipositive ions and tripositive ions of the first transition series, zinc and gallium

| Configuration | $M^{2+}$ | $M^{3+}$ |
|---------------|----------|----------|
| $[Ar]3d^1$ | $^aSc^{2+}$ | $Ti^{3+}$ |
| $[Ar]3d^2$ | $Ti^{2+}$ | $V^{3+}$ |
| $[Ar]3d^3$ | $V^{2+}$ | $Cr^{3+}$ |
| $[Ar]3d^4$ | $Cr^{2+}$ | $Mn^{3+}$ |
| $[Ar]3d^5$ | $Mn^{2+}$ | $Fe^{3+}$ |
| $[Ar]3d^6$ | $Fe^{2+}$ | $Co^{3+}$ |
| $[Ar]3d^7$ | $Co^{2+}$ | $Ni^{3+}$ |
| $[Ar]3d^8$ | $Ni^{2+}$ | $Cu^{3+}$ |
| $[Ar]3d^9$ | $Cu^{2+}$ | — |
| $[Ar]3d^{10}$ | $Zn^{2+}$ | $Ga^{3+}$ |

a   Compounds of scandium(II) are very rare.

One of the characteristic features of the chemistry of the transition elements is the formation of a vast number of complexes such as $[Ti(H_2O)_6]^{3+}$, $Ni(CO)_4$ and $[CoCl(NH_3)_4(H_2O)]^{2+}$. These are molecules that consist of a central metal atom or ion, to which is bonded a number of molecules or ions by coordinate-covalent bonds. We refer to the latter as **ligands**, and the number of electron pairs donated to the metal is its **coordination number**.

⬤ What is the coordination number of the metal in the following complexes?
    (a) $[Ti(H_2O)_6]^{3+}$; (b) $Ni(CO)_4$; (c) $[CoCl(NH_3)_4(H_2O)]^{2+}$.

⬤ (a) 6; (b) 4; (c) 6.

---

[†] This is the energy term that is responsible for the 'special stability' of filled and half-filled shells.

Possibly the most striking property of transition-metal complexes is the wide range of colours they exhibit. This tells us that part of the visible region of the electromagnetic spectrum is being absorbed by the molecule. But what energy changes are actually occurring at the molecular level? There are also intriguing variations in the magnetic behaviour of transition-metal complexes. For example, although they both contain central $Fe^{2+}$ ions, $[Fe(H_2O)_6]^{2+}$ is *paramagnetic* (it is attracted into a magnetic field), but $[Fe(CN)_6]^{4-}$ is *diamagnetic* (it is weakly repelled by a magnetic field). In this book, we shall look at some bonding theories to help us explain these, and other, observations.

Our starting point is crystal-field theory (Section 2). This is a delightfully simple approach, which, provides us with a remarkable insight into the chemical and physical properties of complexes of d-block metals. However, there are cases where this model is inadequate, and where molecular orbital theory is more appropriate. In developing a theory of bonding in transition-metal complexes, our starting point is a consideration of the properties of the d orbitals on the metal ion.

## 1.1  What do d orbitals look like?

There are five d orbitals, which, with reference to a set of mutually perpendicular axes, may be represented by their **boundary surfaces**, the contours inside which a d electron is found 95 per cent of the time. The orbitals shown in Figure 1.1 are strictly those for an electron in a hydrogen atom, but those for electrons in

**Figure 1.1**   The shapes and orientation of the 3d orbitals. Note that in each case the orbital is viewed from the front, so the coordinate axes vary from orbital to orbital.

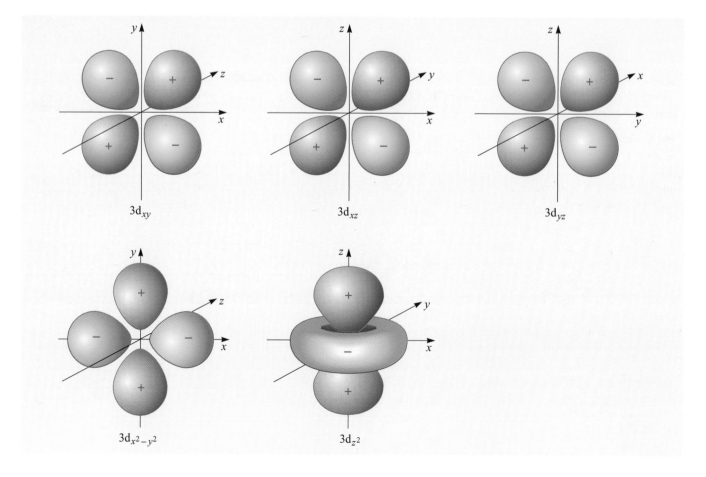

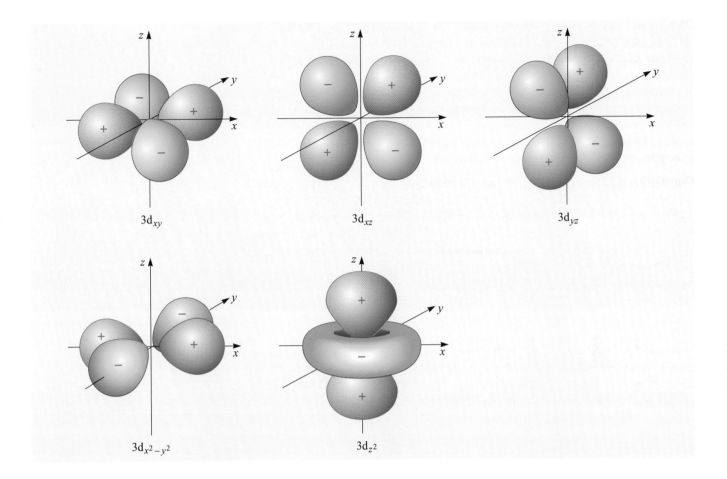

transition-metal ions will have the same shape. In Figure 1.2, the 3d orbitals are all presented from the same perspective, such that the $xz$ plane is always the plane of the paper.

Four of these orbitals have the same shape but are orientated in different directions: the $3d_{xy}$, $3d_{yz}$ and $3d_{xz}$ orbitals have their lobes between the relevant coordinate axes, whereas the $3d_{x^2-y^2}$ orbital has its lobes along the $x$ and $y$ axes. The fifth, $3d_{z^2}$, looks different, but is, in fact, a combination of two orbitals $3d_{y^2-z^2}$ and $3d_{z^2-x^2}$, which are shaped like the other four (Figure 1.3).

**Figure 1.2**   The shapes and orientation of the 3d orbitals, all shown with respect to the $xz$ plane.

**Figure 1.3**   The components of $3d_{z^2}$: the $3d_{y^2-z^2}$ and $3d_{z^2-x^2}$ orbitals.

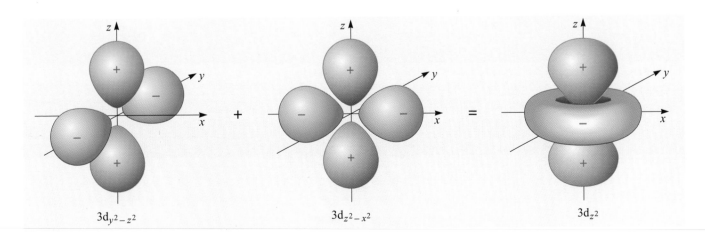

The energies of electrons in the 3d orbitals in transition metals are larger than that of an electron in a 3d orbital in the hydrogen atom. (3d ionisation energies for the first transition series metals are in the range $12.8–17.6 \times 10^{-19}$ J as opposed to $2.4 \times 10^{-19}$ J for hydrogen.) In addition, the energy of the 3d orbital changes when going from the free ion or atom to a complex. It is the changes in the energies of the d orbitals when we add ligands to a 'naked' transition-metal ion that concern us here. How does the energy of a 3d orbital change when a transition-metal ion is surrounded by ligands, and what are the consequences of this change?

# CRYSTAL-FIELD THEORY

We begin our consideration of bonding in transition-metal complexes by looking at crystal-field theory, which is relatively straightforward to apply, and allows us to rationalise, and make predictions about many properties of these molecules.

**Crystal-field theory**, developed by Hans Bethe and John Van Vleck in the 1930s, assumes that ligands behave as point negative charges, and that the metal–ligand interaction occurs on several levels. Overall, a complex will be stabilised relative to the free ion, due to the attraction between the negatively charged ligands and the positively charged metal ion. However, if we take a closer look at the electrons in the metal-ion d orbitals, we would expect *their* energy to increase due to repulsion by the ligands. In other words, the energy of the metal-ion d orbitals will rise. However, this is not the whole story. Given that we are considering an electrostatic interaction, whose magnitude will depend on the distance between the charge centres, we also need to look at how close the d electrons are to the ligands. This, in turn, will depend on which d orbital the electron occupies.

We shall start by looking at the application of crystal-field theory to octahedral complexes, since this geometry is one of the most common in transition-metal chemistry. Our emphasis is on complexes of the first transition series.

## 2.1 Octahedral complexes

We begin by assuming that the ligand negative charges are concentrated at six points representing six octahedrally arranged ligands, two on the $x$-axis, two on the $y$-axis and two on the $z$-axis (Figure 2.1). For a free ion, the d orbitals are energetically equivalent. We already know that the energies of all the d orbitals will increase, but the key question is: are **all** *the 3d orbitals equally affected by this charge?*

To answer this question, let us look at the two d orbitals that are orientated in the $xy$-plane. Figure 2.2 shows the $3d_{xy}$ and $3d_{x^2-y^2}$ orbitals in the $xy$-plane, and the point charges on the $x$- and $y$-axes. By taking this bird's-eye view down the $z$-axis, you can see that whereas the $3d_{x^2-y^2}$ lobes are concentrated towards the point charges, those of the $3d_{xy}$ orbital lie between the charges.

**Figure 2.1**  Six octahedrally disposed ligands represented as point negative charges.

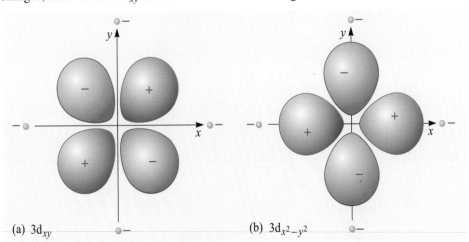

(a) $3d_{xy}$

(b) $3d_{x^2-y^2}$

**Figure 2.2**  (a) $3d_{xy}$ and (b) $3d_{x^2-y^2}$ orbitals surrounded by point negative charges.

**5**

⬤ How do you think this will affect the energy of the two orbitals?

⬤ An electron in a $3d_{x^2-y^2}$ orbital comes closer to the point charges on average than does an electron in a $3d_{xy}$ orbital. Thus, the $3d_{x^2-y^2}$ electron will be repelled more by the ligands, and hence the $3d_{x^2-y^2}$ orbital will be higher in energy than the $3d_{xy}$ orbital.

Similarly, if you look at the $xz$-plane, you will find that an electron in $3d_{z^2}$ will experience a greater repulsion than one in the $3d_{xz}$ orbital, and if you considered the $yz$-plane, you would find that an electron in $3d_{z^2}$ would be repelled more than one in the $3d_{yz}$ orbital.

To summarise: for a set of octahedrally arranged charges (an octahedral crystal field), the energy of the orbitals aligned along the axes ($3d_{x^2-y^2}$ and $3d_{z^2}$) will be higher than those of the $3d_{xy}$, $3d_{xz}$ and $3d_{yz}$ orbitals, which are aligned between the axes (that is, further away from the ligands). This is represented in the form of an energy-level diagram in Figure 2.3. There are several points to note about this diagram. In both Figure 2.3a and b, the five d orbitals all have the same energy (they are referred to as being **degenerate**), and (b) simply represents the average energy of the orbitals in the complex, known as the **barycentre.** This level would correspond to a hypothetical situation in which the metal ion was surrounded by a sphere of negative charge. The splitting of the orbitals is shown in Figure 2.3c; they are 'balanced' about the barycentre. Furthermore, in an octahedral complex, the $3d_{xy}$, $3d_{xz}$ and $3d_{yz}$ orbitals are energetically equivalent, as are the $3d_{z^2}$ and $3d_{x^2-y^2}$ orbitals. The symbol $\Delta_o$ (pronounced delta 'oh' for octahedral) denotes the energy separation between the two sets of orbitals, and is referred to as the **crystal-field splitting energy.** The $3d_{xy}$, $3d_{xz}$ and $3d_{yz}$ orbitals have an energy $\frac{2}{5}\Delta_o$ less than the average energy of the orbitals, and the $3d_{z^2}$ and $3d_{x^2-y^2}$ orbitals are raised $\frac{3}{5}\Delta_o$ higher than the average.

Note that the levels in Figure 2.3 are labelled $t_{2g}$ and $e_g$. These are symmetry labels for a complex (or molecule) belonging to the symmetry point group of an octahedron, $\mathbf{O_h}$.

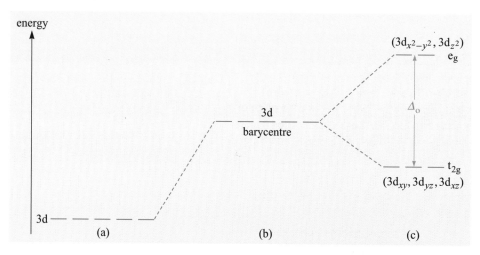

**Figure 2.3** Partial orbital energy-level diagram (showing 3d levels only) for (a) a free transition-metal ion, (b) a transition-metal ion in a sphere of negative charge, and (c) a transition-metal ion in an octahedron of six point negative charges.

We shall consider these symbols in more detail in Section 10 in the context of molecular orbital theory, but for the present you need only consider them as labels:

- t denotes a triply degenerate orbital.
- e denotes a doubly degenerate orbital.

Later you will meet the symmetry labels 'a' and 'b', which are singly degenerate levels. The symbols 'g' and 'u' refer to the behaviour of an orbital under the operation of inversion (p. 57). They are only used for complexes that possess a centre of symmetry.

So we now have an energy-level diagram. But how can it be used to explain the properties of transition-metal complexes? The following steps will get us started:

(i)   determine the oxidation state of the metal ion in the complex;

(ii)  calculate the corresponding number of d electrons;

(iii) establish how these electrons occupy the energy-level diagram (bearing in mind that each energy level can hold a maximum of two electrons).

Firstly, let's consider a complex of titanium in its +3 oxidation state, where there is one d electron ($Ti^{3+}$, $3d^1$). This will enter the $t_{2g}$ level (Figure 2.4a). It does not matter whether we place the electron in the $d_{xy}$, $d_{yz}$ or $d_{xz}$ orbital because they are degenerate. For complexes containing metal ions of configuration $d^2$ ($Ti^{2+}$ and $V^{3+}$) or $d^3$ ($V^{2+}$ and $Cr^{3+}$), the electrons enter the $t_{2g}$ level, but they occupy separate orbitals with parallel spins (Figure 2.4b and c).

The energy of an orbital is determined by the attraction of the nuclei in the complex for an electron in that orbital. Generally, when assigning electrons to orbitals, we ignore any interaction between the electrons. However, in this case, we need to consider the repulsion of the negatively charged electrons in more detail. Two electrons in one orbital will repel each other more than two electrons in different orbitals because, on average, they will be closer together. In addition, electrons with paired spins repel each other more than those with parallel spins. Consequently, electron repulsion is minimised if the electrons are in different orbitals with parallel spins. The energy required to force two electrons into the same orbital is the **pairing energy, $P$**.

The pairing energy becomes important when we reach the $3d^4$ situation, as we are now faced with two choices. The fourth electron could either enter the $t_{2g}$ level and pair with an existing electron, or, it could avoid paying the price of the pairing energy by occupying the $e_g$ level. Which of these possibilities occurs depends on the relative magnitude of the crystal-field splitting and the pairing energy. The two options are:

(i)   If $\Delta_o < P$, the fourth electron goes into the $e_g$ level, with a spin parallel to those of the $t_{2g}$ electrons. This is known as the **weak-field** or **high-spin** case, and is represented by the notation $t_{2g}^3 e_g^1$.

(ii)  If $\Delta_o > P$, the energy required for an electron to occupy the upper level, $e_g$, will outweigh the effect of electron–electron repulsion. The fourth electron then goes into a $t_{2g}$ orbital, where it has to be spin paired. This is represented by the notation $t_{2g}^4 e_g^0$, and is known as the **strong-field** or **low-spin** case.

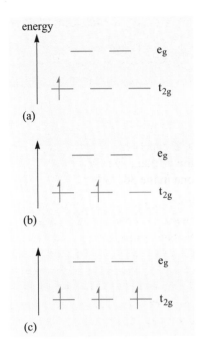

**Figure 2.4**   Occupation of 3d orbitals for (a) $d^1$; (b) $d^2$; (c) $d^3$ ions.

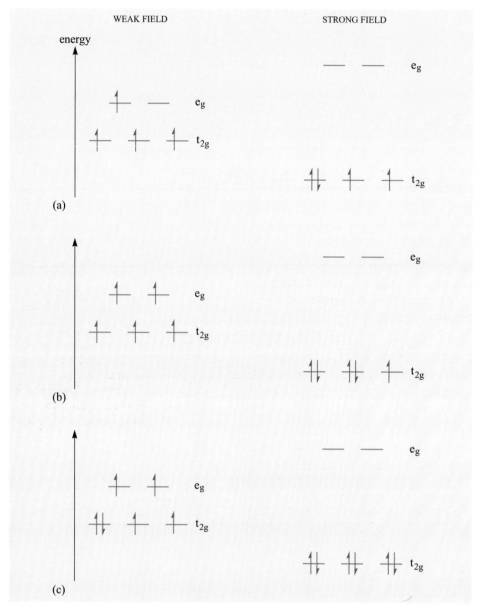

WEEK FIELD        STRONG FIELD

**Figure 2.5** Occupation of 3d orbitals for (a) $d^4$, (b) $d^5$ and (c) $d^6$ complexes in weak and strong octahedral crystal fields.

These two possibilities are considered in Figure 2.5, which also includes high- and low-spin arrangements for $d^5$ and $d^6$.

Sketch orbital energy-level diagrams similar to Figure 2.5 showing the weak-field and strong-field configurations for a $d^7$ complex.

See Figure 2.6.

Would we expect to see high-spin and low-spin complexes for $d^8$ and $d^9$ complexes in an octahedral crystal field?

No; there is only one possible arrangement of electrons in both cases, the size of $\Delta_o$ makes no difference to the occupation of the levels, $d^8$: $t_{2g}^6 e_g^2$ and $d^9$: $t_{2g}^6 e_g^3$ (see Figure 2.7)

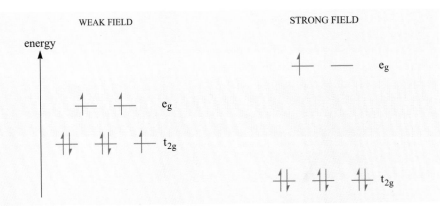

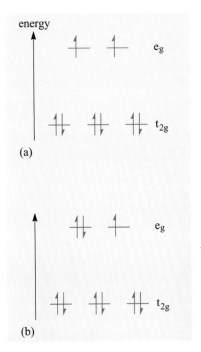

**Figure 2.6** Occupation of 3d orbitals for a $d^7$ complex in weak and strong octahedral crystal fields.

You will notice that the different arrangements of electrons in the $t_{2g}$ and $e_g$ orbitals for $d^4$–$d^7$ ions can result in configurations where the d electrons are completely paired, or contain one or more unpaired electrons; that is, complexes containing these ions can be either diamagnetic or paramagnetic. The magnetic properties of transition-metal complexes are considered in more detail in Section 6, and at this stage we simply note the existence of these two possibilities.

We shall now use these d-electron configurations to explain the variations in an important property of transition-metal ions.

**Figure 2.7** Occupation of 3d orbitals for (a) $d^8$ and (b) $d^9$ complexes in both weak and strong octahedral crystal fields. Note that $\Delta_o$ is bigger for strong-field ligands.

## 2.2 Ionic radii

A plot of the ionic radii of the dipositive ions for the first transition series shows an overall decrease with increasing atomic number, but with a double-bowl shaped profile. This is shown in Figure 2.8. As the nuclear charge increases, electrons enter the same sub-shell (3d); that is, the electrons are roughly the same distance from the nucleus. Electrons in the same shell do not screen the positive charge of the nucleus from each other very effectively. Hence the net nuclear charge experienced by the electrons increases as the atomic number increases. This increased charge causes the electrons to move closer to the nucleus, and hence the ionic radii of the first-row transition elements exhibit an overall decrease across the series.

If there were a spherical distribution of electric charge over the ions, we would expect a regular decrease in ionic radii (shown by the light green line in Figure 2.8), but clearly this is not the case. In fact, taking into account the regular distribution of

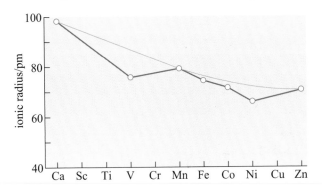

**Figure 2.8** Ionic radii of the divalent ions of calcium and the first-row transition metals in the difluorides.

the d orbitals, this situation would only be achieved for $d^0$, $d^5$ (high spin) and $d^{10}$. Here crystal-field theory can help us. The radii in Figure 2.8 were obtained by measuring metal–fluorine distances in metal difluorides, and the crystal structures of these compounds are such that each metal ion is surrounded by an octahedron of fluoride ions. Hence, to a first approximation we are still dealing with octahedral complexes, so our d-orbital energy-level diagram derived in Section 2.1 (Figure 2.3) will apply.

Let us start with $TiF_2$, which is actually unknown, but we can still use Figure 2.8 to estimate and discuss its notional *internuclear distance*. The $Ti^{2+}$ ion has two 3d electrons, and in an octahedral crystal field they will occupy $t_{2g}$ orbitals with parallel spins. The electrons on the transition metal screen the fluoride ions (which we are regarding as point negative charges) from the charge of the metal nucleus. As the electrons are dividing their time between orbitals that are concentrated between the ligands, we would expect this screening to be less efficient than if the electrons were in $e_g$ orbitals. Since the fluoride ions are screened less than we would expect if the crystal field were spherical, they move closer to the metal nucleus, hence shortening the metal–fluorine distance. Thus, from crystal-field theory, we expect $Ti^{2+}$ to have a smaller ionic radius than if it were a spherical ion.

⬤ What variation from the spherical ion depiction does crystal-field theory predict for $V^{2+}$ in $VF_2$?

⬤ $V^{2+}$ would have three electrons in the $t_{2g}$ level. So, like $Ti^{2+}$, $V^{2+}$ will have a smaller ionic radius than would be expected for a spherical ion; in fact, it has the largest deviation from the spherical ion prediction of all the elements in the first transition series.

For $Cr^{2+}$, there are four d electrons, so we need to consider the possibility of high- and low-spin configurations. However, as we shall see later, fluoride ions are very weak-field ligands, so the complex will be high spin; that is, the fourth electron goes into the $e_g$ level.

The $Cr^{2+}$ ion will therefore have a smaller radius[†] than expected, but the deviation is less than for $V^{2+}$ because we now have electron density in a d orbital pointing directly at the ligands and therefore screening the nuclear charge more efficiently.

⬤ Use crystal-field theory to explain why the ionic radius for $Mn^{2+}$ is that expected for a spherical ion.

⬤ $Mn^{2+}$ is a $d^5$ ion, and in the fluoride has three electrons in $t_{2g}$ and two in $e_g$. Thus, all five d orbitals are equally occupied, and the $Mn^{2+}$ ion is spherical.

Going on to $Fe^{2+}$, $Co^{2+}$ and $Ni^{2+}$, the $t_{2g}$ level is gradually filled, and, like $Ti^{2+}$ and $V^{2+}$, these ions are smaller than expected. $Cu^{2+}$ has its ninth electron in the $e_g$ level, but this level is still not full, and therefore the radius is also less than expected for a spherical ion. $Zn^{2+}$ is on the light green line because it has a filled 3d sub-shell and therefore is a spherical ion.

Crystal-field theory thus helps us understand the double-bowl variation of ionic radii across the first transition series. In the next Section, we shall see how it can be used to explain the variation of other properties.

[†] The plot in Figure 2.8 does not include points for Cr and Cu. The reason for this is that chromium(II) fluoride and copper(II) fluoride have distorted octahedral structures. Why this is so will become apparent in Section 4.

## QUESTION 2.1

The cyanide ion, $CN^-$, gives rise to strong-field complexes. Suppose the ionic radii of the first transition series $M^{2+}$ ions were taken not from fluorides but from cyanides, $M(CN)_2$, in which $M^{2+}$ is octahedrally coordinated by six $CN^-$ ligands. Describe how you would expect the plot of ionic radii to differ from Figure 2.8, and why. Sketch the curve you would predict from crystal-field theory for the ionic radii of the first transition series metal ions in cyanides, $M(CN)_2$.

## QUESTION 2.2

Sketch the predicted variation across the first-row transition-metal series of the ionic radius of $M^{2+}$ in the oxide MO. ($O^{2-}$ is a weak-field ligand, so the ions will have high spin.)

# 2.3 Crystal-field stabilisation energy

We now turn our attention to the variation of the *lattice energy* of the chlorides $MCl_2$, where M is a metal of the first transition series. These are plotted in Figure 2.9, which also has a double-bowl shape. This is not surprising as lattice energy depends on ionic radius, which, as you will recall from Figure 2.8, exhibits a similar variation. Again, we can account for this variation with reference to crystal-field theory.

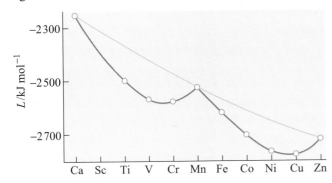

**Figure 2.9** Lattice energies, $L$, of the dichlorides of calcium and the first-row transition metals.

In the dichlorides, the metal ions are in octahedral sites, and for $TiCl_2$, the ion $Ti^{2+}$ has two d electrons, which will enter the $t_{2g}$ orbitals with parallel spins. In Section 2.1 we noted that transition-element electrons in octahedral complexes enter a $t_{2g}$ orbital with an energy $\frac{2}{5}\Delta_o$ less than the barycentre, but those in an $e_g$ orbital are increased in energy by $\frac{3}{5}\Delta_o$. The orbital energy for $Ti^{2+}$ in $TiCl_2$ is thus $2 \times \frac{2}{5}\Delta_o$ or $\frac{4}{5}\Delta_o$ below what we would expect for the ion if it were in a (hypothetical) spherical crystal field. This decrease in energy on going from the spherical situation to an octahedral crystal field is called the **crystal-field stabilisation energy, CFSE** [†].

- What is the CFSE for $V^{2+}$ in an octahedral site?

- $V^{2+}$ has three 3d electrons, and so these will all go into the $t_{2g}$ orbitals with parallel spins. The CFSE for $V^{2+}$ is thus $(3 \times \frac{2}{5})\Delta_o = \frac{6}{5}\Delta_o$.

Returning to our consideration of lattice energies, it is clear that the value for $VCl_2$ in Figure 2.9 is further below the curve for an ion in a spherical environment than that for $TiCl_2$, due to the additional contribution from the CFSE. However, it is important to realise that the magnitude of the CFSE is small (about 10%) compared

[†] Make sure you do not confuse this term with 'crystal-field splitting energy'.

with the total lattice energy. Nevertheless, CFSE does make its presence felt as we move across a transition series.

Now, turning to $CrCl_2$, the problem of weak or strong field has to be considered.

● What is the CFSE for $Cr^{2+}$ in a *weak* octahedral field?

● In the weak-field case, you saw that a $3d^4$ ion such as $Cr^{2+}$ has three electrons in $t_{2g}$ and one in $e_g$ (configuration $t_{2g}^3 e_g^1$). This gives a CFSE of $\frac{6}{5}\Delta_o - \frac{3}{5}\Delta_o = \frac{3}{5}\Delta_o$.

● What is the CFSE for $Cr^{2+}$ in a *strong* octahedral field?

● In a strong-field case, all four electrons go into the $t_{2g}$ orbitals, and so the CFSE is $4 \times \frac{2}{5}\Delta_o = \frac{8}{5}\Delta_o$.

However, the orbital energy is not the only factor to be taken into account in the strong-field case. The pairing energy will act to reduce the CFSE, and so we need to *subtract P for each pair of spins additional to those paired in the free ion.* Therefore, the total CFSE for $Cr^{2+}$ in a strong-field case is thus $\frac{8}{5}\Delta_o - P$.

● What are the CFSEs for strong- and weak-field $Mn^{2+}$ ($3d^5$)?

● In a strong field, $Mn^{2+}$ has all five electrons in $t_{2g}$, with two electron pairs, so the CFSE will be $2\Delta_o - 2P$ ($5 \times \frac{2}{5}\Delta_o - 2P$). In the weak-field case, $Mn^{2+}$ has the configuration $t_{2g}^3 e_g^2$ where all the spins are parallel. The CFSE is thus zero. The CFSEs for $d^1$–$d^{10}$ configurations in weak and strong fields are given in Table 2.1.

**Table 2.1** CFSEs for first transition series ions in weak and strong octahedral fields

| Configuration | CFSE (weak field) | CFSE (strong field) |
|:---:|:---:|:---:|
| $d^1$ | $\frac{2}{5}\Delta_o$ | $\frac{2}{5}\Delta_o$ |
| $d^2$ | $\frac{4}{5}\Delta_o$ | $\frac{4}{5}\Delta_o$ |
| $d^3$ | $\frac{6}{5}\Delta_o$ | $\frac{6}{5}\Delta_o$ |
| $d^4$ | $\frac{3}{5}\Delta_o$ | $\frac{8}{5}\Delta_o - P$ |
| $d^5$ | 0 | $2\Delta_o - 2P$ |
| $d^6$ | $\frac{2}{5}\Delta_o$ | $\frac{12}{5}\Delta_o - 2P$ |
| $d^7$ | $\frac{4}{5}\Delta_o$ | $\frac{9}{5}\Delta_o - P$ |
| $d^8$ | $\frac{6}{5}\Delta_o$ | $\frac{6}{5}\Delta_o$ |
| $d^9$ | $\frac{3}{5}\Delta_o$ | $\frac{3}{5}\Delta_o$ |
| $d^{10}$ | 0 | 0 |

Like the fluoride ion, the chloride ion produces a weak field, and the lattice energies in Figure 2.9 are all for high-spin complexes. You can see that the deviation from the spherical environment curve (light green line) increases from $Ti^{2+}$ to $V^{2+}$, then decreases at $Cr^{2+}$. $Mn^{2+}$ lies on the curve, as it has zero CFSE. From $Fe^{2+}$ through to $Zn^{2+}$ (filled d sub-shell) the pattern is repeated, as first the $t_{2g}$ and then the $e_g$ levels are filled.

## QUESTION 2.3

Draw a rough sketch of how you would expect the lattice energies of the trifluorides to vary along the series $ScF_3$, $TiF_3$, $VF_3$, $CrF_3$, $MnF_3$, $FeF_3$, $CoF_3$, $NiF_3$, $GaF_3$.

## QUESTION 2.4

Sketch the variation in the lattice energy of the dicyanides $M(CN)_2$ across the first transition series (assume $P$ is about $\frac{1}{2}\Delta_o$ for $CN^-$, and that the $M^{2+}$ ions occupy octahedral sites).

# ELECTRONIC SPECTRA OF OCTAHEDRAL COMPLEXES

**3**

As we have already mentioned, one of the most distinctive properties of transition-metal complexes is their wide range of colours. This means that some of the visible spectrum is being removed from white light as it passes through the sample, so the light that emerges is no longer white. The colour of the complex is complementary to that which is absorbed. The *complementary colour* is the colour generated from the wavelengths left over; for example, if green light is absorbed by the complex, it appears red.

The complex $[Ti(H_2O)_6]^{3+}$ absorbs yellow–green light. What colour does it appear?

Violet.

But, what are the origins of these light-absorbing processes?

The colours result from **electronic transitions** between partially filled d orbitals, which, as we have seen, are split into a lower-energy $\mathbf{t_{2g}}$ **set**, and higher-energy $\mathbf{e_g}$ **set**. These are called d $\leftrightarrow$ d or d–d (spoken as 'dee dee') transitions, and the colour we observe is thus a measure of the crystal-field splitting, $\Delta_o$. Indeed, transition-metal ions are not coloured in the gaseous state, so the splitting of the d orbitals is crucial for these colours in the solid state and in solution. Figure 3.1 shows the electronic spectrum of $[Ti(H_2O)_6]^{3+}$. The horizontal scale is wavenumber, for which the common symbol is $\sigma$, which is measured in $cm^{-1}$.

The peak in Figure 3.1 is due to an electronic transition from $t_{2g}$ to $e_g$ in the $[Ti(H_2O)_6]^{3+}$ ion, [†] and may be represented by the energy-level diagram shown in the inset. This transition confers the violet colour on the complex. Note that

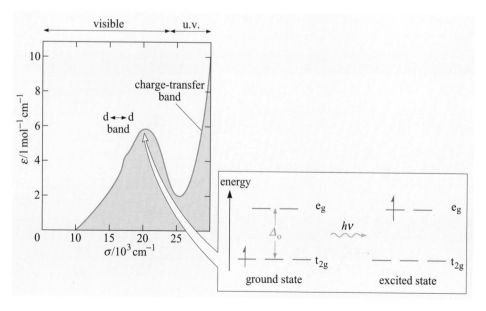

**Figure 3.1** Electronic absorption spectrum of $[Ti(H_2O)_6]^{3+}$; the inset shows the transition of an electron from the $t_{2g}$ level to the $e_g$ level, which gives rise to the d–d band. (The unit of molar absorption coefficient is litre $mol^{-1}$ $cm^{-1}$, but in this text we shall abbreviate it to l $mol^{-1}$ $cm^{-1}$.)

[†] You will also notice a much more intense transition, at higher energy in these spectra. This is a *charge-transfer band*, which in some cases may encroach into the visible region, producing intensely coloured complexes. The origin of these bands is considered in Section 14.

$61\,mol^{-1}\,cm^{-1}$ represents a low value of the molar absorption coefficient, which is a characteristic feature of **d–d spectra**. In contrast, if you have ever recorded the electronic spectra of organic compounds such as dyes, you will have observed molar absorption coefficients in excess of $10\,000$ times the values of d–d bands.

Measuring the wavenumber of maximum absorption for complexes with one 3d electron as Figure 3.1 should therefore give us a value of $\Delta_o$. This spectrum is very broad, which is a feature frequently observed in d–d spectra. The width of d–d bands is due to metal–ligand vibrations, which result in a range of $\Delta$ values corresponding to the range of metal–ligand bond distances encompassed by vibrational motion. (This is analogous to the rotational fine structure of vibrational spectra.)

We have already seen that transition-metal complexes exhibit a variety of colours, which are dictated by the value of $\Delta_o$. It would therefore be of interest to examine more closely the factors that affect the crystal-field splitting.

Figure 3.2 (p. 16) shows schematic spectra for titanium(III) complexes involving different ligands. The peak corresponding to the d–d transition clearly occurs at different energies.

● With reference to Figures 3.1 and 3.2, arrange the ligands $H_2O$, $F^-$, urea, $Cl^-$, $Br^-$ and oxalate (ox, $(COO^-)_2$) in order of increasing $\Delta_o$ — in other words, going from the weakest-field to the strongest-field ligand.

● The larger the value of $\Delta_o$, the more energy is absorbed in the transition from $t_{2g}$ to $e_g$, and the higher the wavenumber of the spectral band. The order here is thus:

$$Br^- < Cl^- < urea < F^- < H_2O < ox$$

A series like this, which arranges ligands in order of their $\Delta_o$, is known as a **spectrochemical series**. By studying a large number of complexes, it is possible to compile a series that holds for complexes of these ligands with most transition metals. A more extensive series is given in the margin; ligands towards the end of the series such as $PR_3$, $CN^-$ and CO are strong-field ligands.

Note that $SCN^-$ represents a thiocyanate anion bound to a metal through S, and $NCS^-$ the same anion bound through N. In $PR_3$, R is used to represent a saturated alkyl group. The ligands $edta^{4-}$ (ethylenediaminetetraacetate), en (ethylenediamine), bipy (2,2′-bipyridyl) and phen (1,10-phenanthroline) are shown in Structures **3.1** to **3.4**; the coordinating atoms are shown in green.

CH$_2$−N(CH$_2$COO$^-$)$_2$
|
CH$_2$−N(CH$_2$COO$^-$)$_2$

**3.1** edta$^{4-}$

H$_2$N        NH$_2$
    \        /
    CH$_2$−CH$_2$

**3.2** en

**3.3** bipy

**3.4** phen

The spectrochemical series (margin):

increasing $\Delta_o$

WEAK FIELD: I$^-$
Br$^-$
SCN$^-$
Cl$^-$
S$^{2-}$
F$^-$
urea ≈ OH$^-$
ox ≈ O$^{2-}$
H$_2$O
NCS$^-$
edta$^{4-}$
NH$_3$
en
bipy ≈ phen ≈ NO$_2^-$
PR$_3$
CN$^-$
STRONG FIELD: CO

The spectrochemical series

**Figure 3.2**  Schematic electronic absorption spectra of (a) $[TiF_6]^{3-}$, (b) $[Ti[(urea)_6]^{3+}$, (c) $TiCl_3$, (d) $TiBr_3$, (e) $[Ti(ox)_3]^{3-}$.

The relative order of ligands in the spectrochemical series is considered in more detail in Section 9, but at this stage you should be able to *roughly* link the series with the position of the coordinating atom in the Periodic Table: crystal-field strength decreases from Group IV (14) to Group VII (17). For the moment, this should be a useful *aide memoire*.

The influence of the ligand on the colour of a complex may be illustrated by considering the $[Ni(H_2O)_6]^{2+}$ complex, which forms when nickel(II) chloride is dissolved in water. If the bidentate ligand $^\dagger$ ethylenediamine (en) is progressively added in the molar ratios en : Ni, 1 : 1, 2 : 1, 3 : 1, the following series of reactions and their associated colour changes occur:

$$[Ni(H_2O)_6]^{2+}(aq) + en(aq) = [Ni(H_2O)_4en]^{2+}(aq) + 2H_2O(l) \qquad (3.1)$$

   green                            pale blue

$$[Ni(H_2O)_4en]^{2+}(aq) + en(aq) = [Ni(H_2O)_2en_2]^{2+}(aq) + 2H_2O(l) \qquad (3.2)$$

                         blue/purple

$$[Ni(H_2O)_2en_2]^{2+}(aq) + en(aq) = [Nien_3]^{2+}(aq) + 2H_2O(l) \qquad (3.3)$$

                        violet

This sequence is shown in Figure 3.3.

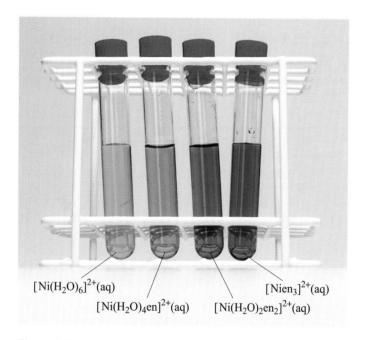

[Ni(H_2O)_6]^{2+}(aq)          [Nien_3]^{2+}(aq)
    [Ni(H_2O)_4en]^{2+}(aq)   [Ni(H_2O)_2en_2]^{2+}(aq)

**Figure 3.3**  Aqueous solutions of complexes of nickel(II) with an increasing number of ethylenediamine ligands.

It is interesting to stop and think about this spectrochemical series for a minute. In particular, how well does it tie in with the purely electrostatic (repulsive) treatment, which forms the basis of crystal-field theory? By this approach we

$^\dagger$A ligand containing two coordinating atoms.

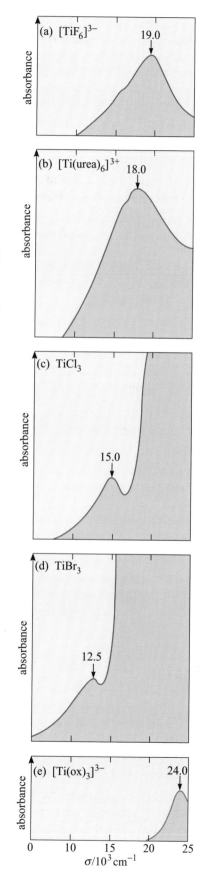

(a) $[TiF_6]^{3-}$    19.0

(b) $[Ti(urea)_6]^{3+}$   18.0

(c) $TiCl_3$    15.0

(d) $TiBr_3$    12.5

(e) $[Ti(ox)_3]^{3-}$    24.0

absorbance

$\sigma/10^3\,cm^{-1}$

would expect more highly charged ligands to produce stronger fields, and consequently larger values of $\Delta_o$. However, the charged halides produce very weak fields, whereas the strongest field is produced by CO, a neutral molecule! Even water produces a stronger field than its charged relatives $OH^-$ and $O^{2-}$. Thus, crystal-field theory gives no indication as to why ligands occupy the positions they do in the series. To account for this we need to consider covalent bonding in the metal–ligand interaction. This is where molecular orbital theory fits in (Sections 8–14). However, for the time being we will stick with the crystal-field approach, which successfully accounts for many of the properties of complexes.

We have seen that the size of $\Delta_o$ depends on the nature of the ligand, but there are also two ways in which the central metal ion can contribute:

(i) For a given ligand and a given metal, crystal-field splitting increases with increase in metal oxidation state; for example, $\Delta_o$ is $9\,400\ cm^{-1}$ for $[Fe^{II}(H_2O)_6]^{2+}$, but $13\,700\ cm^{-1}$ for $[Fe^{III}(H_2O)_6]^{3+}$. On simple electrostatic grounds, we can argue that the increase in the positive charge on the metal draws the ligands closer, which results in greater repulsion between the metal d electrons and the ligand point charges, hence causing an increase in $\Delta_o$.

(ii) For a given ligand, crystal-field splitting increases on descending a Group in the Periodic Table. This is demonstrated in Table 3.1 for Co, Rh and Ir. In this case, we could conjecture that, as the size of the d orbital increases on descending the Group (that is, Co 3d, Rh 4d, Ir 5d), the d electrons are progressively closer to the ligands, and so $\Delta_o$ increases.

**Table 3.1** Crystal-field splitting for ammine complexes of the Co, Rh and Ir Group

| Complex | $\Delta_o/cm^{-1}$ |
|---|---|
| $[Co(NH_3)_6]^{3+}$ | 23 000 |
| $[Rh(NH_3)_6]^{3+}$ | 34 000 |
| $[Ir(NH_3)_6]^{3+}$ | 41 000 |

So far, we have only considered transitions of one electron from $t_{2g} \rightarrow e_g$, which produce a single absorption band. This is fine for a $d^1$ complex, but a representative spectrum of a manganese(II) complex (Figure 3.4) reveals a more complicated situation.

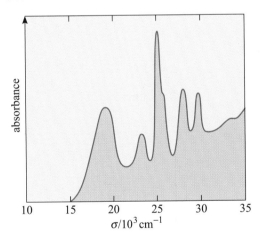

**Figure 3.4** Schematic electronic absorption spectrum of $[Mn(H_2O)_6]^{2+}$.

The value of $\Delta_o$ for $[Mn(H_2O)_6]^{2+}$ cannot be obtained simply from this spectrum as it contains several d–d bands. This is because when there is more than one electron, we need to take into account not only the orbital energy change, but also the change in repulsion energy between the d electrons when a transition occurs.

It is not a simple matter to identify the transitions giving rise to d–d bands for complexes with more than one d electron, but a consideration of the $d^2$ case will give you an idea of how the electron repulsion terms affect the spectra. A $d^2$ ion will have bands corresponding to one of its d electrons undergoing a transition from $t_{2g}$ to $e_g$.

In other words, we start with two electrons in $t_{2g}$ and end up with one in the $t_{2g}$ and one in the $e_g$ level. Supposing that the electron remaining in the $t_{2g}$ level is in a $d_{xy}$ orbital, then the $e_g$ electron can be in $d_{x^2-y^2}$ or $d_{z^2}$. (Note that we refer to the different way electrons can occupy the orbitals in a particular configuration as a **state**.)

Would the repulsion between an electron in $d_{xy}$ and one in $d_{x^2-y^2}$ be the same as that between one in $d_{xy}$ and one in $d_{z^2}$?

No. The electrons in $d_{xy}$ and $d_{x^2-y^2}$ will be closer to each other on average than one in $d_{xy}$ and one in $d_{z^2}$. Therefore, we would expect a greater repulsion between an electron in $d_{xy}$ and one in $d_{x^2-y^2}$.

A different amount of energy would therefore be needed for a transition from $t_{2g}$ to $d_{x^2-y^2}$ than that for a transition from $t_{2g}$ to $d_{z^2}$, and so we would expect to see two bands instead of one for the $t_{2g} \rightarrow e_g$ transition. Neither band would correspond exactly to $\Delta_0$. The $[V(H_2O)_6]^{3+}$ ion, for example, has three bands, comprising two bands around $17\,800\ cm^{-1}$ and one band at $25\,700\ cm^{-1}$, corresponding to a transition of one electron from $t_{2g}$ to $e_g$, whereas $\Delta_0$ for this complex is $16\,900\ cm^{-1}$.

So our explanation of $d^2$ spectra is still over-simplified.

Can we be sure that the electron remaining in $t_{2g}$ is in $d_{xy}$?

No. It could be in $d_{yz}$ or $d_{xz}$ or $d_{xy}$.

But whichever $t_{2g}$ orbital the electron occupies, there is still a difference in repulsion energy depending on which orbital the $e_g$ electron occupies. So, although our assumption that the electron is in $d_{xy}$ was arbitrary, our conclusion that there will be more than one energy for the configuration $t_{2g}^1 e_g^1$ is still valid.

With more than two electrons, the spectra become even more complicated. For example, manganese(II) complexes have five d electrons, and even for a simple complex such as $[Mn(H_2O)_6]^{2+}$, there are seven d–d bands, six of which are apparent in the spectrum shown in Figure 3.4.

How a little colour from a transition-metal impurity can confer considerable added value to humble materials like alumina is the subject of Box 3.1.

As we have seen, d–d bands in the visible spectrum are very weak; that is, they have small molar absorption coefficients, $\varepsilon$. Just as in rotational and vibrational spectroscopy, there are selection rules which determine if a particular transition is 'allowed'. By simple analogy with everyday life, we can view these selection rules rather like a set of traffic lights. When at red, vehicles must stop, but, we know the occasional driver will (either inadvertently or deliberately) pass through! In essence, selection rules dictate the probability of a transition occurring. Similarly, even though particular d–d bands are formally forbidden, there are mechanisms that can account for the occasional photon being absorbed, imparting a weak colour to the complex.

The **Laporte selection rule** requires that during an electronic transition the orbital quantum number[†], $l$, can only change by $\pm 1$. In other words, we can have transitions from an s orbital to a p orbital, $p \rightarrow s$, $p \rightarrow d$, etc., but not $s \rightarrow d$. Indeed the $d \leftrightarrow d$ bands we have discussed previously are also forbidden. So if the Laporte selection rule were truly obeyed, many transition-metal complexes would *not* be coloured!

---

[†] This is sometimes referred to as the 'second' or 'azimuthal' quantum number.

## Box 3.1 Gemstones – from dross to jewels

The colours produced by electronic transitions within the d orbitals of a transition-metal ion occur frequently in everyday life. One example is provided by the much-cherished colours of many gemstones, which consist of transition-metal ions incorporated into a normally white mineral. Aluminium oxide, one form of which is the mineral corundum (or $\alpha$-alumina), has a crystal structure that may be described as a close-packed array of oxide ions, with $Al^{3+}$ ions occupying two-thirds of the octahedral holes.

Ruby (Figure 3.5a) is aluminium oxide ($Al_2O_3$) containing about 0.5–1% $Cr^{3+}$ ions ($d^3$), which are randomly distributed in positions normally occupied by $Al^{3+}$. We may view these chromium(III) species as octahedral chromium(III) complexes incorporated into the alumina lattice; d–d transitions at these centres give rise to the colour. There are two main transitions, one in the yellow–green region, and the other closer to violet; overall this confers the characteristic brilliant red colour to ruby.

In emerald (Figure 3.5b), $Cr^{3+}$ ions occupy octahedral sites in the mineral beryl ($Be_3Al_2Si_6O_{18}$). However, in this case the dimensions of the octahedra are larger than in aluminium oxide, and hence the crystal-field splitting is smaller. Consequently, the absorption bands seen in ruby shift to longer wavelength, namely yellow–red and blue, causing emerald to transmit light in the green region.

(a)

(b)

**Figure 3.5** (a) Ruby: this gemstone was found in marble from Mogok, Burma; crystal length, 12 mm; (b) emerald: this gemstone was found in Muzo, Columbia; crystal length, 10 mm.

In complexes, there is some relaxation of this rule, due to the mixing of ligand character with the metal 3d orbitals (this is a feature of the molecular orbital approach to bonding, which will be developed in Sections 8–14). One feature that remains, though, is that if a complex has a *centre of symmetry* (Section 10.1), a transition from a g orbital to another g orbital, or from a u orbital to another u orbital, is forbidden. Thus, the transition from $t_{2g}$ to $e_g$ is forbidden, because both orbitals have the same symmetry with respect to inversion through the centre of symmetry.

Electronic transitions must also obey the **spin selection rule**. This states that an electron cannot change its spin while undergoing a transition to another level. Considering a d–d transition of a $d^5$ ion, it is clear that if an electron is to be excited from $t_{2g}$ to $e_g$, it must change its spin as there is no room in the upper level for an electron with the same spin (Figure 3.6). Such a transition is **spin forbidden**.

**Figure 3.6** A spin-forbidden d–d transition for a $d^5$ ion in a weak-field octahedral complex.

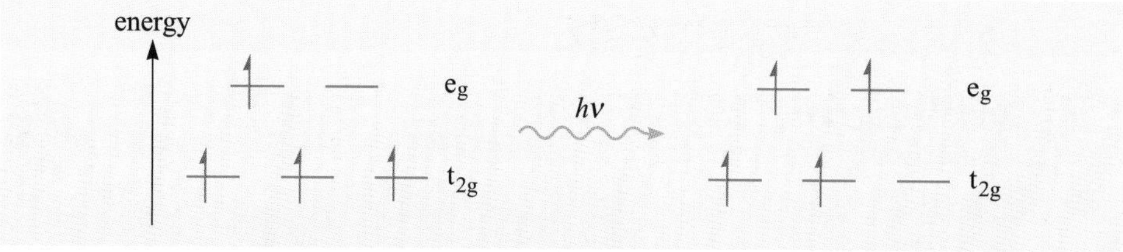

energy

$e_g$    $hv$    $e_g$

$t_{2g}$    $t_{2g}$

**Figure 3.7** A spin-allowed d–d transition for a $d^4$ ion in a weak-field octahedral complex.

However, all d–d transitions are not spin forbidden. For example, in a $d^4$ ion an electron can be excited from $t_{2g}$ to $e_g$ without changing its spin (Figure 3.7).

To see what this means in practice, we could take an equimolar solution of the complexes $[Co(H_2O)_6]^{2+}$ and $[Mn(H_2O)_6]^{2+}$. Although the former is not strongly coloured, the latter is virtually colourless, and appears white as a powdered solid. For a high-spin $d^7$ ion like $Co^{2+}$, an electron can be promoted from a $t_{2g}$ orbital to an $e_g$ orbital without breaking the spin selection rule (it still breaks the Laporte rule though). However, $Mn^{2+}$ is a $d^5$ ion, and, as shown in Figure 3.6, promotion is impossible in a high-spin environment without reversing the electron spin in the excited state, so in this case d–d transitions are forbidden by both the Laporte and spin selection rules.

One question that may have occurred to you is that, since the d–d transitions in a complex such as $[Mn(H_2O)_6]^{2+}$ are forbidden both by the Laporte and spin selection rules, why do we see them at all? Well, even transitions such as these become partly allowed if we take vibrations into account. For example, certain molecular vibrations may transiently remove the centre of symmetry (Figure 3.8). Hence an electronic transition becomes partially allowed. This is known as **vibronic** (**vib**rational–elect**ronic**) **coupling**.

## QUESTION 3.1

A large number of cobalt complexes have been investigated. $\Delta_o$ values obtained from the electronic spectra of some octahedral cobalt(III) complexes are given in Table 3.2. Arrange the ligands in a spectrochemical series, and show that this order is consistent with the general series given on p. 15.

**Table 3.2** $\Delta_o$ values for some cobalt(III) complexes

| Ligand | $\Delta_o$/cm$^{-1}$ |
| --- | --- |
| $F^-$ | 13 100 |
| ox | 18 000 |
| $CNO^-$ | 26 100 |
| $CN^-$ | 32 200 |
| $S_2PF_2^-$ | 14 240 |

## QUESTION 3.2

Do the d–d electronic transition(s) responsible for the green colour of $[Ni(H_2O)_6]^{2+}(aq)$ obey the Laporte and spin selection rules?

## QUESTION 3.3

If you look at an empty milkbottle, you can frequently see a green tint in the glass. This is particularly apparent if the bottle is viewed along its axis, rather than side on. Given that $Fe^{3+}$ is a common impurity in glass, how would you account for the weak colour?

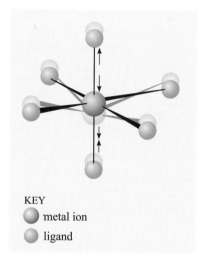

KEY
- metal ion
- ligand

**Figure 3.8** An asymmetric vibration in an octahedral complex, leading to the removal of the centre of symmetry (defined on p. 57). The positions of the atoms at two stages of the asymmetric vibration are superimposed; dark colours are used for the first positions of the atoms, and light colours for the later positions. In the first position there is a centre of symmetry, but this is not present in the second. Movement of one pair of ligand atoms is clarified by arrows.

# SUBSTITUTED AND DISTORTED OCTAHEDRAL COMPLEXES, AND SQUARE-PLANAR COMPLEXES

4

So far, we have only dealt with regular octahedral complexes, in which the metal ion is surrounded by six equidistant identical ligand atoms. Although convenient for teaching the principles of crystal-field theory, such complexes are very much the exception. Complexes are found with other geometries, such as tetrahedral or square planar, and even nominal octahedral complexes may contain two different ligands, for example $[TiCl_2(H_2O)_4]^+$, or show other irregularities, which means that they do not belong to the $O_h$ symmetry point group [†].

We begin by considering complexes that are not truly octahedral, which means the symbols $t_{2g}$ and $e_g$ are no longer appropriate labels for the orbital energy levels. As we shall see, this may cause the 3d levels to split in different ways. The first group of complexes all belong to the symmetry point group $D_{4h}$:

- complexes of the type $[MA_4B_2]^{n\pm}$, where two A ligands *trans* to each other in the octahedron have been substituted by B ligands (Structure **4.1**);

- distorted octahedral complexes $[ML_6]^{n\pm}$, in which two *trans* metal–ligand distances are shorter or longer than the other four (Structure **4.2**);

- square-planar complexes (Structure **4.3**).

Firstly, we shall consider what happens when an octahedral complex is distorted by gradually removing two ligands *trans* to each other, and slightly shortening the metal–ligand distances of the other four, thereby maintaining the average energy of the orbitals.

**4.1**

**4.2**

**4.3**

Suppose that the ligands to be removed are on the *z*-axis. What will be the effect on the energy of the $3d_{z^2}$ orbital of the metal ion as these ligands move further out?

The repulsion between the electrons in the $3d_{z^2}$ orbital and the ligands will decrease, which will result in the energy of the $3d_{z^2}$ orbital being lowered.

Will the energy of the $3d_{x^2-y^2}$ orbital be similarly affected?

No, an electron in $3d_{x^2-y^2}$ is repelled by the ligands on the *x*- and *y*- axes, and as these have moved in slightly, the energy of the $3d_{x^2-y^2}$ orbital is raised.

Thus, the two $e_g$ orbitals (the $3d_{z^2}$ and $3d_{x^2-y^2}$) in the octahedral complex, are now *no longer at the same energy*; that is, their degeneracy has been lifted, and the $3d_{z^2}$ will be at lower energy than the $3d_{x^2-y^2}$. How about the other three d orbitals? The $3d_{xy}$ (like $3d_{x^2-y^2}$) will still be repelled by the now closer ligands on the *x*- and *y*-axes, but $3d_{xz}$ and $3d_{yz}$ will be lowered in energy because the electrons occupying them will experience less repulsion from the more-distant ligands on the *z*-axis.

[†] Note that molecular symmetry and symmetry point groups are discussed in more detail in Section 10.

So, overall, the $t_{2g}$ level in this distorted octahedral complex will now split into two levels, which are labelled $e_g$ ($d_{xz}$, $d_{yz}$) and $b_{2g}$ ($d_{xy}$). The original $e_g$ level also splits into two: $a_{1g}$ ($d_{z^2}$) and $b_{1g}$ ($d_{x^2-y^2}$), as shown in the middle column of Figure 4.1. Note that because the $t_{2g}$ orbitals do not point directly at the ligands, the splitting of the $t_{2g}$ level is much less than that of the $e_g$ level.

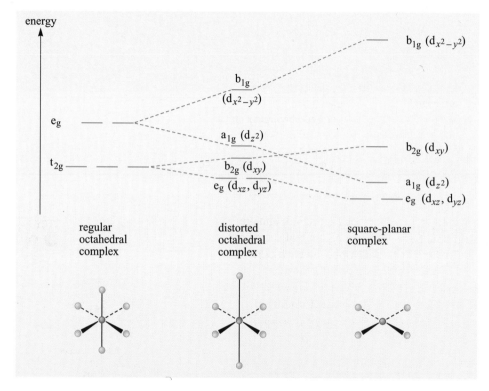

**Figure 4.1** Splitting of the 3d levels for a regular octahedral complex, a distorted octahedral complex (with two *trans* bonds longer than the other four), and a square-planar complex.

As the ligands move further out, the splitting of the octahedral levels increases. Again, the effect is more marked for the $d_{z^2}$ and $d_{x^2-y^2}$ orbitals because they are directed towards the ligands.

If the *trans* ligands are removed altogether, we then have a square-planar complex. As illustrated in the right-hand column of Figure 4.1, the $3d_{z^2}$ level ($a_{1g}$) may fall below that of $3d_{xy}$ ($b_{2g}$).

Now let's consider the situation in which the two *trans* ligands are moved closer to the metal ion, and the four in the $xy$ plane are moved further out. In this case, the *trans* ligands will repel an electron in the $3d_{z^2}$ orbital more than they would in a regular octahedral complex.

- What effect will this have on the $3d_{z^2}$ orbital energy?

- The $3d_{z^2}$ orbital will be even higher in energy than in the regular octahedral complex.

As with the previous example, the $e_g$ level is split into two, but this time the higher level is $a_{1g}$ ($d_{z^2}$) and the lower level is $b_{1g}$ ($d_{x^2-y^2}$). The $t_{2g}$ level will also split. This time, the higher level will be $e_g$ ($d_{xz}$, $d_{yz}$) and the lower level will be $b_{2g}$ ($d_{xy}$). The orbital energy-level diagram for this type of complex is shown in Figure 4.2.

Now, what sort of compounds are found to adopt the geometries discussed above?

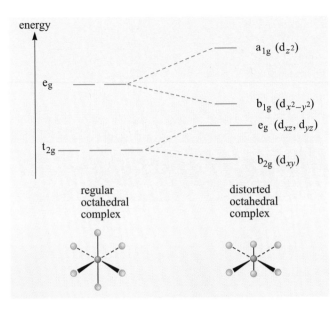

**Figure 4.2** Splitting of the 3d levels for a regular octahedral complex and a distorted octahedral complex in which two *trans* ligands on the $z$-axis are closer to the metal ion than the four in the $xy$ plane.

One group of examples are the halides of copper(II), such as Figure 4.3, and the chromium(II) halides, which in the solid state have distorted octahedral structures, where the metal ion is surrounded by two elongated bonds opposite each other (*trans*), and four shorter bonds in the $xy$ plane: two ligands are further away than the other four, giving a tetragonal distortion; these complexes are said to be **Jahn–Teller distorted**. The **Jahn–Teller theorem**, which was formulated in 1937 by H. A. Jahn and E. Teller, is applicable to any non-linear molecule, although it has proved particularly relevant to transition-metal complexes.

> The theorem says that a non-linear molecule is unstable in a degenerate state and will distort to remove the degeneracy.

To understand what this means, we need to think again about what we mean by a 'state'. To illustrate this, let us take a high-spin (weak-field) complex of chromium(II), such as $[Cr(H_2O)_6]^{2+}$.

🔵 How many 3d electrons does the $Cr^{2+}$ ion have?

🔵 Four.

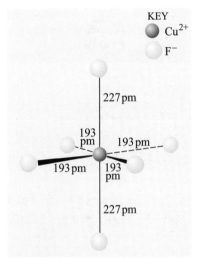

**Figure 4.3** An example of Jahn–Teller distortion in copper(II) fluoride.

In a weak-field octahedral complex, these electrons will go into the three $t_{2g}$ orbitals and one $e_g$ orbital with parallel spins. Now, the $t_{2g}$ electrons can only be arranged one way: one of them has to be in $d_{xy}$, one in $d_{xz}$ and one in $d_{yz}$. But the $e_g$ electron can be in either $d_{z^2}$ or $d_{x^2-y^2}$. In a regular octahedral complex, the energy of the complex would be the same for either of these alternatives, and so the complex would be in either of two degenerate states.

🔵 Deviation from octahedral symmetry can also arise by removing the degeneracy of the $t_{2g}$ set. In this case, which other high-spin (weak field) configurations would you expect to be degenerate in an octahedral field?

🔵 Complexes of titanium(III) and vanadium(III) ($d^1$ and $d^2$, respectively) will be degenerate because there is a choice of $t_{2g}$ orbitals to occupy. Iron(II) and cobalt(II) complexes ($d^6$ and $d^7$, respectively) also have a choice of $t_{2g}$ orbitals available, and so they too are degenerate.

**23**

Copper(II) complexes ($d^9$) are degenerate because like the $d^4$ case there is a choice of $e_g$ orbital.

Why should complexes having such electronic configurations distort? Well, consider the complexes of copper(II) and chromium(II), which are in a degenerate state due to partial occupation of the $e_g$ orbitals. It was noted earlier that complexes of these ions are often distorted by having two ligands further away than the other four.

⬤ What happens to the $e_g$ level when an octahedral complex is distorted in this way?

⬤ The $e_g$ level splits into two: $a_{1g}$ ($d_{z^2}$) and $b_{1g}$ ($d_{x^2-y^2}$).

*If the total electrostatic interaction between ligands and metal ions remains the same* (that is, the distorted complex is formed by moving four ligands nearer to the metal and two ligands further away from the metal), the $a_{1g}$ level must be lower in energy and the $b_{1g}$ higher in energy than the $e_g$ level in the regular octahedral complex (Figure 4.1).

⬤ In a high-spin $d^4$ complex, which levels do the electrons occupy if the complex is Jahn–Teller distorted as we have indicated?

⬤ The electrons will go into the $e_g$ ($d_{xz}$, $d_{yz}$) and, in order to avoid pairing, $b_{2g}$ ($d_{xy}$) from the octahedral $t_{2g}$, and into $a_{1g}$ ($d_{z^2}$) from the octahedral $e_g$ (Figure 4.4).

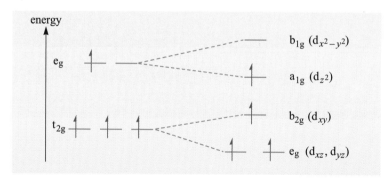

**Figure 4.4** Occupation of the $t_{2g}$ and $e_g$ levels for a $d^4$ Jahn–Teller distorted octahedral complex.

Since $a_{1g}$ is lower in energy than the regular octahedral $e_g$ level, the distorted complex is more favourable energetically than a regular octahedral complex; that is, the highest-energy electron is in a lower energy state than if it were in an undistorted octahedral complex.

⬤ Does this explanation hold for copper(II) complexes as well?

⬤ Yes. For copper(II), two electrons will go into the $a_{1g}$ ($d_{z^2}$) and one into $b_{1g}$, compared with three in $e_g$ for the regular octahedral complex.

Complexes of chromium(II) and copper(II) thus distort because there is an *overall reduction* in orbital energy to be achieved on doing so.

● Despite the fact that they may exist in degenerate electronic states, the distortions from regular octahedral geometry observed in complexes with $d^1$, $d^2$, $d^4$ (low spin), $d^5$ (low spin), $d^6$ (high spin) and $d^7$ (high spin) are relatively small in comparison with $d^4$ (high spin) and $d^9$ complexes. What explanation can you suggest for this?

● For $d^1$, $d^2$, $d^4$ (low spin), $d^5$ (low spin), $d^6$ (high spin) and $d^7$ (high spin), the degenerate state is the $t_{2g}$ set of metal orbitals. Because $t_{2g}$ orbitals are not directed towards the ligands, the splitting of the $t_{2g}$ orbitals in the distorted state is much smaller.

As shown in Figure 4.2, some complexes exist with two metal–ligand distances shorter than the other four (that is, compressed down the $z$-axis). For example, in the compound $K_2CuF_4$, there are two Cu—F distances of 195 pm and four Cu—F distances of 208 pm. In this case, the $e_g$ level, will again split into two, but this time the $b_{1g}$ level will be lower in energy than the regular octahedral $e_g$, and the $a_{1g}$ level will be higher in energy. The complex will have two electrons in $b_{1g}$ and one in $a_{1g}$, which, according to the Jahn–Teller theorem, again gives the distorted state an energy advantage over the regular octahedral situation with three electrons in $e_g$ (Figure 4.5).

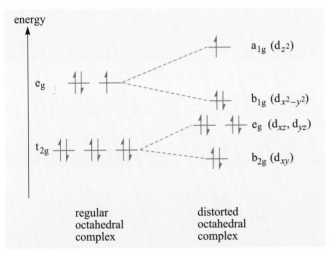

**Figure 4.5**  Splitting of the 3d levels for the distorted octahedral complex $K_2CuF_4$, in which two *trans* ligands on the $z$-axis are closer to the copper(II) ion than the four in the $xy$ plane.

The $Cu^{2+}$ ($d^9$) ion is nearly always found in a distorted ligand environment. Even cases where an apparently regular octahedral geometry has been reported can be accounted for by a *dynamic* Jahn–Teller effect, in which the direction of elongation is constantly changing. One of the few exceptions is $Cu^{2+}$ in copper(II) hexafluoro-silicate, $[Cu(H_2O)_6]SiF_6$. Even here, however, X-ray crystallographic studies have revealed that only one-quarter of the copper ions occupy sites where all the Cu—O distances are equal; the remainder occupy considerably distorted sites.

The Jahn–Teller effect applies equally well to excited states as it does to ground-state d-electron configurations, as can be seen from the electronic spectra of appropriate complexes. If you look back at Figure 3.1, you will notice a shoulder on the d–d band of the electronic spectrum $[Ti(H_2O)_6]^{3+}$. This is a consequence of two closely spaced absorption bands overlapping, which arise from electronic transitions from the $d_{xz}$, $d_{yz}$, $d_{xy}$ levels to $a_{1g}$ and $b_{1g}$ levels. In other words, the excited state of this ion is a degenerate electronic state, which is subject to Jahn–Teller splitting into two components. This is shown schematically (in a rather exaggerated form) in Figure 4.6. A more striking example is given in Figure 4.7, which shows the absorption spectrum of $K_2Na[CoF_6]$, which contains the complex ion $[CoF_6]^{3-}$ ($Co^{III}$, $d^6$): two component bands, arising from the Jahn–Teller distortion of the $t_{2g}^3 e_g^3$ excited state, are clearly defined.

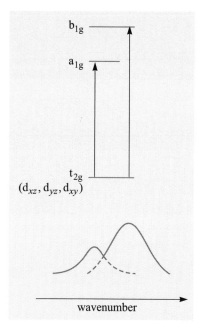

**Figure 4.6** Energy-level diagram showing the electronic transitions in $[Ti(H_2O)_6]^{3+}$, taking into account splitting of the $e_g$ level, and the corresponding part of the electronic spectrum. For simplicity, the splitting of the $t_{2g}$ level is not shown.

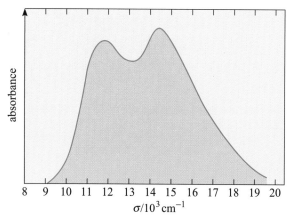

**Figure 4.7** The electronic absorption spectrum of the $[CoF_6]^{3-}$ ion, showing the two peaks due to the Jahn–Teller splitting of the excited state ($t_{2g}^3 e_g^3$).

*Square-planar complexes* are particularly common at the end of the transition-metal series, especially for ions with a $d^8$ and $d^9$ configuration. Palladium and platinum, for example, form many square-planar complexes. One notable example is *cis-*$[PtCl_2(NH_3)_2)]$, commonly known as *cisplatin* (Structure **4.4**), a potent anti-cancer drug, which has been widely used in chemotherapy. Referring to the right-hand column of Figure 4.1, we can begin to see why this geometry is favoured by $d^8$ and $d^9$ ions. Platinum(II) has eight d electrons, with the configuration $t_{2g}^6 e_g^2$ in a regular octahedral complex. In a square-planar complex with strong-field ligands, the gap between $b_{1g}$ and $b_{2g}$ is large, and eight electrons fill the $e_g$, $a_{1g}$ and $b_{2g}$ levels, leaving the $b_{1g}$ level empty. This means that the highest-occupied energy level is

$$H_3N \diagdown \quad \diagup Cl$$
$$Pt$$
$$H_3N \diagup \quad \diagdown Cl$$

**4.4**

lower in energy than it would have been in an octahedral complex. Thus, there is an energy advantage in forming a square-planar complex. Of course, we have to balance this against other factors, such as the pairing of the two electrons in $b_{2g}$ and the different number of metal–ligand bonds in the two geometries. Thus, we expect square-planar complexes with strong-field ligands, where the gain in orbital energy is sufficient to offset these other terms, and for second-row and third-row transition elements such as palladium and platinum, for which $\Delta$ will be larger than for first-row transition elements.

Nickel(II), which has the configuration $3d^8$, forms square-planar complexes with strong-field ligands such as cyanide ($CN^-$). With weak-field ligands such as halide ions and ligands coordinating through N and O, nickel(II) forms octahedral complexes such as $[NiF_6]^{4-}$ and $[Ni(NH_3)_6]^{2+}$. (However, see Box 4.1.) Four-coordinate complexes of nickel(II) with weak-field ligands are tetrahedral (for example, $[NiCl_4]^{2-}$).

### QUESTION 4.1

Which of the following high-spin complexes would you expect to exhibit a Jahn–Teller distortion?

(a) $[Cr(NH_3)_6]^{2+}$; (b) $[MnCl_6]^{3-}$; (c) $[Fe(H_2O)_6]^{3+}$.

## Box 4.1 The mystery of the Lifschitz salts

In certain cases, complexes of nickel(II) can be produced which contain a mixture of square-planar and octahedral molecules, or the coordination geometry can be subtly manipulated. Classic examples are the so-called Lifschitz salts (named after I. Lifschitz who first prepared them). These are complexes of nickel(II) with substituted ethylenediamines (an example is shown as Structure **4.5**). What confused chemists for many years, was that sometimes these complexes were blue and paramagnetic, or, alternatively, they were yellow and diamagnetic, depending on factors such as: the structure of the diamine, the nature of the anion, the solvent used during preparation, temperature and exposure to water vapour.

$$(C_6H_5)HC - H_2N \cdots \quad \cdots NH_2 - CH(C_6H_5)$$
$$| \qquad\qquad Ni \qquad\qquad |$$
$$(C_6H_5)HC - H_2N \quad\quad NH_2 - CH(C_6H_5)$$

**4.5**

It has now been demonstrated that the yellow species are square planar, with all electrons spin-paired. This form may be converted to an octahedral complex by coordination of ligands above and below the plane; these ligands could be water, solvent molecules or anions. The octahedral complexes are blue with two unpaired electrons in the $e_g$ level.

# TETRAHEDRAL COMPLEXES

5

Tetrahedral complexes are found quite widely in transition-metal chemistry, and again, the bonding in complexes of this geometry can be successfully treated using crystal-field theory. To see how the 3d levels split in a tetrahedral environment, it is probably easiest to imagine a tetrahedral complex in a cube as in Figure 5.1, with the ligands occupying alternate vertices and the $x$-, $y$-, and $z$-axes through the faces of the cube.

The crucial point to remember when applying crystal-field theory to tetrahedral complexes is that *no d orbital points directly at the ligands* (although, as we shall see, some are closer than others).

⬤ Think, for example, about the $3d_{xy}$ and $3d_{x^2-y^2}$ orbitals. Will electrons in these orbitals be equally repelled by the ligands in a tetrahedral complex?

⬤ No. Although neither orbital points directly towards the ligands, an electron in $3d_{xy}$ (Figure 5.2a) will be closer to the ligands and so will be repelled more than one in $3d_{x^2-y^2}$ (Figure 5.2b).

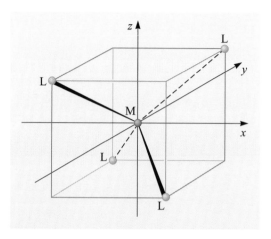

**Figure 5.1**  A tetrahedral complex in a cube, showing $x$-, $y$- and $z$-axes.

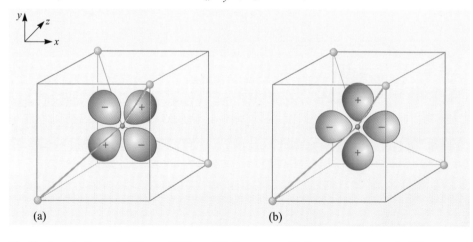

(a)   (b)

**Figure 5.2**  (a) A tetrahedral complex in a cube, showing the disposition of the ligands with respect to the metal $3d_{xy}$ orbital; (b) the same complex showing the position of the $3d_{x^2-y^2}$ orbital.

Similarly, electrons in $3d_{xz}$ and $3d_{yz}$ orbitals are repelled more than one in $3d_{z^2}$. Hence, for a tetrahedral complex, the $3d_{xy}$, $3d_{xz}$ and $3d_{yz}$ orbitals are higher in energy than $3d_{x^2-y^2}$ and $3d_{z^2}$. The orbital energy-level diagram for a tetrahedral complex is shown in Figure 5.3. Like the octahedral case, the energy of *all* the d orbitals is raised relative to the free ion in Figure 5.3a.

The lower level ($3d_{x^2-y^2}$ and $3d_{z^2}$) in Figure 5.3c is labelled **e** and the higher level **t₂**. (Note the absence of a g subscript here; it is not required for tetrahedral complexes because they do not have a centre of symmetry.) The energy gap is labelled $\Delta_t$ (t for tetrahedral). Since none of the d orbitals point directly at the ligands, the difference in energy between e and $t_2$ is not as large as the difference between $t_{2g}$ and $e_g$ in octahedral complexes. In fact, if the metal ion, the ligands and the metal–ligand distance are the same in both octahedral and tetrahedral cases, it can be shown that $\Delta_t \approx \frac{4}{9}\Delta_o$. As a consequence, *virtually all tetrahedral complexes are high spin* due to the smaller crystal-field splitting. There are exceptions, however, as Box 5.1 reveals.

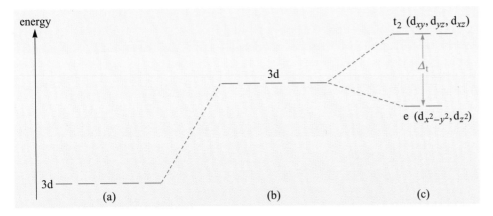

**Figure 5.3** Orbital energy-level diagram (showing 3d orbitals only) for (a) a free ion, (b) an ion in a sphere of negative charge and (c) an ion in a tetrahedral complex.

## Box 5.1 A new kind of tetrahedral complex

In 1986, Klaus Theopold and colleagues at Cornell University, USA reported the first example of a low-spin tetrahedral complex of a first-row transition metal. It was the cobalt–alkyl complex, tetrakis(1-norbornyl)cobalt(IV). A simplified structure is given in Figure 5.4, in which the norbornyl ligand is shown as a seven-carbon-atom skeleton; it may be viewed as a six-membered ring with a bridging carbon atom (see inset). Cobalt was reported to be in the +4 oxidation state, with a $d^5$ configuration. Instead of an expected high-spin arrangement with five unpaired electrons (see top margin diagram), magnetic susceptibility measurements (see Section 6) indicated only one unpaired electron (see bottom margin diagram), consistent with the novel low-spin configuration $e^4 t_2^1$.

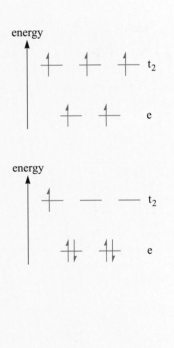

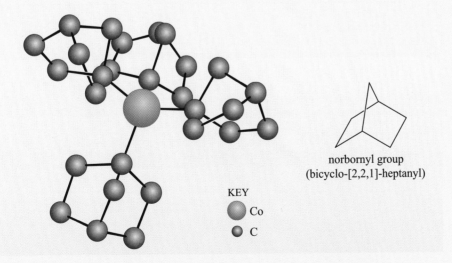

norbornyl group
(bicyclo-[2,2,1]-heptanyl)

KEY
Co
C

**Figure 5.4** A simplified representation of the low-spin tetrahedral complex, tetrakis(1-norbornyl)cobalt(IV); the structural formula of the norbornyl group is shown on the right. For clarity, the hydrogen atoms have been omitted.

The colours of tetrahedral complexes are far more intense than their octahedral counterparts at the same concentration. This is because they do not possess a centre of symmetry, and, consequently, the Laporte selection rule is not strictly applicable. This is discussed in more detail in Section 11 using the molecular orbital approach.

Figure 5.5 shows the effect of adding concentrated hydrochloric acid to an aqueous solution of cobalt(II) chloride. The pale pink solution arises from the octahedral complex $[Co(H_2O)_6]^{2+}$, which is converted to the tetrahedral $[CoCl_4]^{2-}$ on addition of the acid:

$$[Co(H_2O)_6]^{2+}(aq) + 4Cl^-(aq) \rightleftharpoons [CoCl_4]^{2-}(aq) + 6H_2O(l) \qquad (5.1)$$
$$\text{pink} \qquad\qquad\qquad\qquad \text{blue}$$

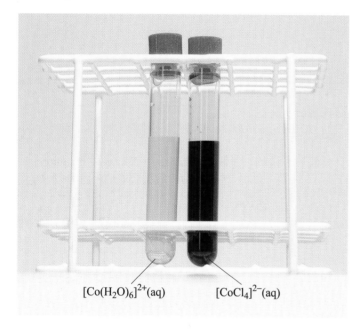

$[Co(H_2O)_6]^{2+}(aq)$      $[CoCl_4]^{2-}(aq)$

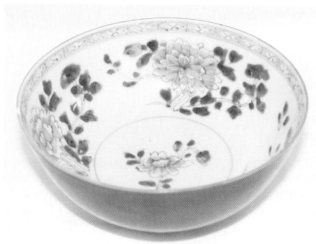

**Figure 5.6** A Chinese bowl from the 18th century, decorated with a cobalt(II) glaze.

**Figure 5.5** The pale pink solution (left-hand tube) was formed by dissolving cobalt(II) chloride in water. On adding concentrated hydrochloric acid, a deep blue solution (right-hand tube) containing the ion $[CoCl_4]^{2-}$ resulted.

A very common decorative colouring in ceramics (Figure 5.6) and glass is the deep blue of $Co^{2+}$ in a tetrahedral environment, and the indicator used in the desiccant silica-gel is a tetrahedral $Co^{2+}$ complex, which on absorbing moisture is converted to a pale-pink octahedral complex.

● In addition to being more strongly coloured, tetrahedral manganese(II) complexes are often green, whereas octahedral $[Mn(H_2O)_6]^{2+}$ is pale pink. What reason can you suggest for this?

● The crystal-field splitting for tetrahedral complexes, $\Delta_t$, is smaller than for octahedral $\Delta_o$. Thus, the bands in the spectra of the tetrahedral complex would be expected to be at lower wavenumber (longer wavelength) — that is, further towards the red end of the visible spectrum — than those in the spectra of the octahedral complexes. A complex that absorbs in the red will appear green. Conversely, octahedral complexes, which absorb towards the green/blue end of the visible spectrum, appear red or, because the absorptions are very weak, pink.

CFSE in a tetrahedral complex may be calculated in exactly the same way as for octahedral complexes. Each electron in the e level contributes $\frac{3}{5}\Delta_t$ and each electron in the $t_2$ level contributes $-\frac{2}{5}\Delta_t$ to the CFSE.

The maxima in the tetrahedral CFSE occur at $d^2$ and $d^7$, which in part explains the occurrence of $V^{3+}$ ($d^2$) tetrahedral complexes such as $VX_4^-$ (where X = Cl, Br, I) and the fact that cobalt(II) ($d^7$) forms more tetrahedral complexes than any other transition-metal ion. This is developed further in Section 5.1. The CFSEs for $d^1$–$d^{10}$ configurations in tetrahedral complexes are given in Table 5.1.

## 5.1 The occurrence of tetrahedral and square-planar four-coordinate complexes

The most common stereochemistry for transition-metal complexes is octahedral, but towards the end of the transition-metal rows we find four-coordinate complexes. For the first row, these are most common in complexes of cobalt(II), nickel(II), copper(II), copper(I) and zinc(II). Table 5.2 shows the most common four-coordinate geometries for complexes of these metals.

In the absence of CFSE, the four ligands will arrange themselves to be as far apart as possible — that is, tetrahedrally — and this is the most common four-coordinate geometry, as in copper(I) with its full d shell ($3d^{10}$). For the $d^7$ and $d^8$ ions cobalt(II) and nickel(II), we saw earlier that square-planar complexes have an energy advantage over octahedral ones for strong-field ligands. For copper(II) there is an energy advantage for square-planar complexes with both strong- and weak-field ligands. If we compare the orbital energy-level diagram for tetrahedral and square-planar $d^8$ complexes (Figure 5.7), we can see that there is a similar advantage for square-planar over tetrahedral complexes. So we would expect square-planar geometry for copper(II) complexes and for low-spin (strong field) cobalt(II) and nickel(II) complexes on crystal-field grounds.

**Table 5.1** CFSEs for first transition series ions in tetrahedral complexes

| Configuration | CFSE |
|:---:|:---:|
| $d^1$ | $\frac{3}{5}\Delta_t$ |
| $d^2$ | $\frac{6}{5}\Delta_t$ |
| $d^3$ | $\frac{4}{5}\Delta_t$ |
| $d^4$ | $\frac{2}{5}\Delta_t$ |
| $d^5$ | 0 |
| $d^6$ | $\frac{3}{5}\Delta_t$ |
| $d^7$ | $\frac{6}{5}\Delta_t$ |
| $d^8$ | $\frac{4}{5}\Delta_t$ |
| $d^9$ | $\frac{2}{5}\Delta_t$ |
| $d^{10}$ | 0 |

**Table 5.2** The most common four-coordinate geometries for first-row transition elements

| Element | Geometry |
|---|---|
| cobalt(II) | tetrahedral |
| nickel(II) | square planar and tetrahedral |
| copper(II) | square planar |
| copper(I) | tetrahedral |
| zinc(II) | tetrahedral |

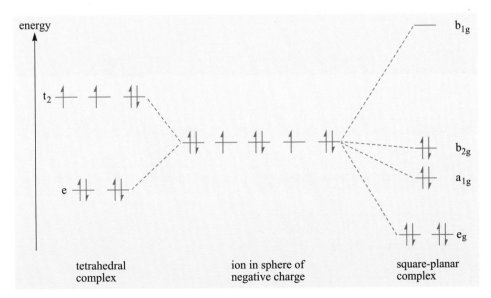

**Figure 5.7** 3d levels in tetrahedral and square-planar $d^8$ complexes.

Other factors also affect the geometry though, particularly steric constraints imposed by multidentate ligands. For example, very large ligands may force a complex to be tetrahedral because the ligands will then be less 'squashed'. On the other hand, the structures of some bidentate ligands, such as ethylenediamine, may be such that the distance between the two coordinating atoms fits a square plane better than a tetrahedron.

It should be noted that no crystal-field splitting labels ($\Delta$) are shown on the d-orbital energy-level diagram for a square-planar complex. This is because there are several possible energy-level differences. However, the difference in energy between the $b_{2g}$ ($d_{xy}$) and $b_{1g}$ ($d_{x^2-y^2}$) is the same as $\Delta_o$ if the complex is made of the same metal and ligands in both cases.

### QUESTION 5.1

How many unpaired electrons are there on cobalt in the tetrahedral complex $[CoCl_4]^{2-}$?

### QUESTION 5.2

(a)   Sketch the predicted variation across the first-row transition series for the enthalpy of reaction of the following reaction:

$$M^{2+}(g) + 4Cl^-(g) = [MCl_4]^{2-}(g) \tag{5.2}$$

($[MCl_4]^{2-}$ is a tetrahedral ion.)

(b)   In Equation 5.2, why is the physical state of the metal ion shown as gaseous?

# MAGNETISM AND THE MAGNETIC PROPERTIES OF TRANSITION-METAL COMPLEXES

6

In general, there are several ways in which a substance can behave in a magnetic field. A magnetic field induces electrons to circulate and repels **diamagnetic substances**. A circulating charge produces a magnetic moment, and this is in the opposite direction to the field, thereby causing repulsion. Diamagnetism occurs in all matter, but it is a weak effect.

In this Section we shall be concerned with **paramagnetic substances**, which are attracted to a magnetic field. Paramagnetism is due to isolated unpaired spins of electrons. If unpaired spins are coupled to each other, then the material can be *ferromagnetic*, *antiferromagnetic* or *ferrimagnetic*. (Ferromagnetism occurs in iron and cobalt, and is the property generally referred to as 'magnetism' in everyday life.)

At this stage, we shall consider isolated complexes only, see how paramagnetism can be measured and how unpaired spins give rise to this property. We can think of magnetic fields as being produced by circulating charge; for example, a magnetic field is produced by an electric current flowing through a solenoid. The magnetic field strength, $H$, can be expressed in terms of the current density in the solenoid. The magnetic field produced can also be described in terms of lines of magnetic force, and the density of these lines gives us the magnetic flux density, $B$. In a vacuum, $B$ and $H$ are related by a constant called the *permeability of free space*, $\mu_0$, which has the value $4\pi \times 10^{-7}\,\text{T m A}^{-1}$:

$$B = \mu_0 H \tag{6.1}$$

If a paramagnetic substance is placed in the magnetic field, it contributes its own field due to circulating charges. The total field strength is then increased by the field strength of the sample, which is called the *magnetisation*, $M$.

⬤ How does this affect the flux density, $B$?

⬤ Since the total field strength has increased, the magnetic flux density also increases.

This can be expressed by Equation 6.2:

$$B = \mu_0(H + M) \tag{6.2}$$

The quantity usually measured is the **magnetic susceptibility**, $\chi$ (pronounced 'kye'). This is defined as follows:

$$\chi = \frac{M}{H} \tag{6.3}$$

so that Equation 6.2 can be rewritten

$$B = \mu_0 H(1 + \chi) \tag{6.4}$$

The magnetic susceptibility can be measured by investigating the change in weight of a sample when a magnetic field is applied. If a paramagnetic sample is placed unsymmetrically in a magnetic field, then the weight of the sample changes due to the magnetic force. The change in weight divided by the weight in the absence of a magnetic field is proportional to the magnetic susceptibility per unit mass, $\chi_w$ (where the subscript w confusingly stands for mass, to distinguish this quantity from molar susceptibility, $\chi_m$; see later). $\chi_w$ is related to the dimensionless magnetic susceptibility, $\chi$, in Equation 6.4 by

$$\chi_w = \frac{\chi}{4\pi\rho} \qquad (6.5)$$

where $\rho$ (pronounced 'rho') is the density of the sample.

Both diamagnetic and paramagnetic susceptibilities are independent of the field strength, $H$.

The magnetic susceptibility, $\chi^{dia}$, for diamagnetic substances is independent of temperature. If we subtract the diamagnetic contribution from the measured magnetic susceptibility (by using published tables, called Pascal's constants, of contributions for all the atoms present), we find that the remaining paramagnetic susceptibility, $\chi^{para}$, does depend on temperature. Experimentally, it was shown by Pierre Curie that the paramagnetic susceptibility is proportional to the inverse of the temperature:

$$\chi^{para} = \frac{C}{T} \qquad (6.6)$$

where $C$ is the **Curie constant**, which is *characteristic of the complex*, and $T$ is temperature. This behaviour is particular to paramagnetic substances.

To see why a paramagnetic sample behaves in this way, it is useful to think of the sample as a collection of transition-metal complex ions, each of which behaves as a tiny magnet. An applied magnetic field tends to make them line up with the field, but when the temperature is above $0\,K$, they also have thermal energy, and this tends to make them move around. This movement alters their orientation so that they are no longer lined up with the field but are more randomly distributed in direction. As the temperature rises, the thermal energy increases and the magnets become more randomly orientated. Thus, as shown by Equation 6.6, the magnetic susceptibility varies, whereas the value of the magnetic field produced by one of these little magnets remains constant with temperature. It is these magnetic fields, known as **magnetic moments**, $\mu$, which give us chemically useful information, and their value can be obtained from the Curie constant, $C$. It is usual to quote the magnetic moment in **Bohr magneton**, $\mu_B$ units; one Bohr magneton has the value $9.274 \times 10^{-24}\,J\,T^{-1}$. The quantity obtained from weight measurements is the susceptibility per unit mass, $\chi_w$. To find the magnetic moment, $\chi_w$ is multiplied by the relative molecular mass of the substance, $M_r$, to give the molar susceptibility, $\chi_m$, which can be corrected for diamagnetic contributions.

$$\chi_m = \chi_w \times M_r \qquad (6.7)$$

The remaining part, $\chi_m{}^{para}$, gives us the magnetic moment via **Curie's law** in the form shown in Equation 6.8:

$$\chi_m{}^{para} = \frac{C'}{T}\mu^2 \qquad (6.8)$$

where $C'$ is a combination of fundamental constants and hence *is the same for all paramagnetic substances*, and $\mu$ is the magnetic moment of the substance being measured. $\chi_m$ for paramagnetic substances is of the order of $10^{-9}$–$10^{-6}\,m^3\,mol^{-1}$ and for diamagnetic substances is $10^{-11}$–$10^{-9}\,m^3\,mol^{-1}$.

One of the simplest and commonly used methods of measuring magnetic susceptibility is the Gouy method. This utilises an accurate balance and a powerful magnet (Figure 6.1). The sample is suspended in such a way that the lower end of the sample is in a region of high field strength, and the upper end is in a region of negligible magnetic field. In this field gradient, a paramagnetic substance will experience a downward force into the field, registering an increase in weight. Conversely, diamagnetic substances show a decrease in recorded weight. Figure 6.2 shows photographs of a modern sophisticated version of this apparatus, known as a *Faraday balance*.

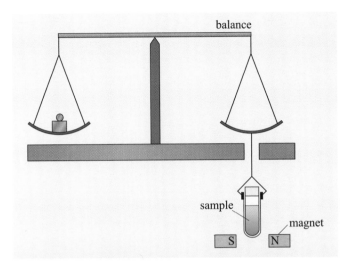

**Figure 6.1** Schematic representation of the Gouy method for determining magnetic susceptibility.

(a)

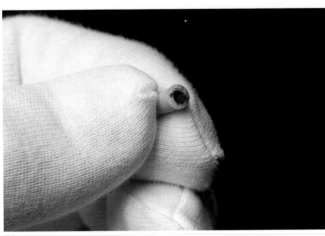

(b)

**Figure 6.2** A Faraday balance. (a) Sample in a Teflon holder suspended between the poles of an electromagnet (notice that the poles are not flat, but curved), which generates the necessary field gradient. (b) A close up of the Teflon holder; typically about 20 mg of sample is used. The blue powder shown here is $Hg[Co(NCS)_4]$, a common reference material for magnetic measurements.

We said that the magnetic moment was produced by unpaired electrons, and now we see how we can relate the magnetic moment to the number of unpaired electrons in a complex. You saw earlier that circulating charges produce a magnetic field.

● An electron is a charged particle. What type of circular motion does a free electron undergo?

● All electrons have spin. Spin in a large object, such as the Earth spinning on its axis, is a circular motion. Although strictly it is not appropriate to think of an electron as a sphere spinning on an axis, electron spin does give rise to a magnetic field.

This field is the **spin magnetic moment**, $\mu_S$. The spin magnetic moment of an electron is quantised; that is, it can only take certain values, depending on the value of the spin quantum number. In Bohr magneton units, the spin magnetic moment is given by Equation 6.9:

$$\mu_S = g\sqrt{S(S+1)}\mu_B \qquad (6.9)$$

where $g$ is a constant known as the *gyromagnetic ratio* or **g factor**, and for a free electron has the value 2.002 3. Because magnetic moments cannot usually be measured so accurately, it is sufficient here to use the value 2 for $g$. $S$ is the total spin quantum number for an atom; for one electron it has the value $\frac{1}{2}$. However, many transition-metal complexes have more than one unpaired electron. Luckily, calculations of $S$ are very simple because each unpaired electron contributes $\frac{1}{2}$, so for two electrons $S = 1$, for three electrons $S = \frac{3}{2}$, and so on.

● What is the maximum number of unpaired spins found in transition-metal complexes?

● Five. There are five d orbitals, and the maximum number of unpaired spins is when there is one electron in each d orbital.

Table 6.1 gives the values of $S$ and $\mu_S$ for 1–5 unpaired electrons.

**Table 6.1**  Values of $S$ and $\mu_S$ for 1–5 unpaired electrons

| Number of unpaired electrons | $S$ | $\mu_S/\mu_B$ |
|:---:|:---:|:---:|
| 1 | $\frac{1}{2}$ | $2\sqrt{\frac{1}{2} \times \frac{3}{2}} = 1.73$ |
| 2 | 1 | $2\sqrt{1 \times 2} = 2.83$ |
| 3 | $\frac{3}{2}$ | $2\sqrt{\frac{3}{2} \times \frac{5}{2}} = 3.87$ |
| 4 | 2 | $2\sqrt{2 \times 3} = 4.90$ |
| 5 | $\frac{5}{2}$ | $2\sqrt{\frac{5}{2} \times \frac{7}{2}} = 5.92$ |

Assuming the magnetic moment is due to electron spin only, it is possible to determine the number of unpaired electrons in a complex from its measured value. In terms of the number of unpaired electrons, $n$, Equation 6.9 can be re-written as

$$\mu_S = \sqrt{n(n+2)}\mu_B \qquad (6.10)$$

For reasons that will become apparent shortly, this equation is known as the 'spin-only' formula'.

How many unpaired electrons would you expect for the complex $[CoF_6]^{3-}$?

This is a $d^6$ high-spin octahedral complex, so will have four unpaired electrons.

Magnetic measurements can be used to identify weak-field and strong-field complexes.

An octahedral cobalt(II) complex ion was found to have a magnetic moment of $1.92\mu_B$. What is the electronic configuration of the metal ion?

Cobalt(II) complexes have seven 3d electrons. In a weak-field octahedral environment, these would give a configuration $t_{2g}^5 e_g^2$, with three unpaired electrons. In a strong field, the configuration would be $t_{2g}^6 e_g^1$, with only one unpaired electron. $1.92\mu_B$ is close to the spin magnetic moment for one unpaired electron, and so we can conclude that this is a strong-field complex.

Table 6.2 shows the observed magnetic moment ($\mu_{obs}$) ranges for octahedral (weak field) and tetrahedral complexes for first-row transition-metal elements in their common oxidation states.

**Table 6.2** Observed magnetic moment ranges for octahedral (weak field) and tetrahedral complexes for first-row transition-metal ions

| Number of d electrons | $\mu_S/\mu_B$ | Octahedral complex | | Tetrahedral complex | |
|---|---|---|---|---|---|
| | | Configuration | $\mu_{obs}/\mu_B$ | Configuration | $\mu_{obs}/\mu_B$ |
| 1 | 1.73 | $t_{2g}^1$ | 1.7–1.8 | $e^1$ | $\approx 1.7$ |
| 2 | 2.83 | $t_{2g}^2$ | 2.8–2.9 | $e^2$ | 2.6–3.0 |
| 3 | 3.87 | $t_{2g}^3$ | 3.7–3.9 | $e^2 t_2^1$ | — |
| 4 | 4.90 | $t_{2g}^3 e_g^1$ | 4.8–5.0 | $e^2 t_2^2$ | — |
| 5 | 5.92 | $t_{2g}^3 e_g^2$ | 5.8–6.0 | $e^2 t_2^3$ | 5.8–6.0 |
| 6 | 4.90 | $t_{2g}^4 e_g^2$ | 5.1–5.7 | $e^3 t_2^3$ | 5.0–5.2 |
| 7 | 3.87 | $t_{2g}^5 e_g^2$ | 4.3–5.2 | $e^4 t_2^3$ | 4.4–4.8 |
| 8 | 2.83 | $t_{2g}^6 e_g^2$ | 2.8–3.4 | $e^4 t_2^4$ | 3.7–4.0 |
| 9 | 1.73 | $t_{2g}^6 e_g^3$ | 1.7–2.2 | $e^4 t_2^5$ | 1.7–2.2 |
| 10 | 0 | $t_{2g}^6 e_g^4$ | 0 | $e^4 t_2^6$ | 0 |

As you can see, many of the values are close to $\mu_S$, but some ions, particularly $d^6$, $d^7$ and tetrahedral $d^8$ ions, have larger magnetic moments than expected. We need to find an explanation for this.

So far, we have only considered contributions to the magnetic moment from the electron spin.

We suggested an analogy for electron spin of the Earth spinning on its axis. What other type of roughly circular motion does the Earth undergo?

The Earth goes round the Sun; this is orbital motion.

An electron in an atom or an ion goes round the nucleus (although we should not think of it as simply travelling round in a fixed orbit like the Earth round the Sun).

Thus, the electron has orbital motion too. This gives rise to an **orbital magnetic moment**. We have to combine the spin and orbital contributions to obtain the total magnetic moment, $\mu_{S+L}$:

$$\mu_{S+L} = \sqrt{g^2 S(S+1) + L(L+1)}\,\mu_B \tag{6.11}$$

where $L$ is the quantum number for an atom which defines the orbital contribution. An electron in a d orbital has an orbital quantum number, $l = 2^\dagger$. For a $d^2$ atom, however, $L$ is the combined quantum number for both electrons. For the spin quantum number, we found $S$ just by adding on $\frac{1}{2}$ for every unpaired electron, but for orbital angular momentum, things are not so simple.

All the unpaired electrons could have $s = \frac{1}{2}$ and $m_s = \pm\frac{1}{2}$ because they would have different values of other quantum numbers$^\dagger$. All d electrons have $l = 2$, but there are five possible values of the magnetic quantum number $m_l$, namely +2, +1, 0, –1 or –2. If there are two d electrons, they must have different values of $m_l$. The value of $L$ is found by taking the maximum value of $m_l$ for the two electrons. The largest value for a single electron is +2, but if one electron has +2, the other electron must have one of the other values. The maximum value left is +1. $L$ is thus $2 + 1 = 3$. The third electron has to have $m_l = 0$ for the maximum value, and so $L = 3$ for $d^3$ as well.

○ What are the values of $L$ for $d^4$ and $d^5$ configurations with all spins unpaired?

○ If all the spins are unpaired, all the electrons must have different values of $m_l$. For $d^4$, $L = 2 + 1 + 0 + (-1) = 2$, and for $d^5$, $L = 2 + 1 + (0) + (-1) + (-2) = 0$.

○ Calculate the total magnetic moment for high-spin ions of configuration $d^1$ to $d^5$.

○ Using Equation 6.11, the following values are obtained:

| No. of unpaired electrons | $S$ | $L$ | $\mu_{S+L}/\mu_B$ |
|---|---|---|---|
| 1 | $\frac{1}{2}$ | 2 | 3 |
| 2 | 1 | 3 | 4.47 |
| 3 | $\frac{3}{2}$ | 3 | 5.20 |
| 4 | 2 | 2 | 5.48 |
| 5 | $\frac{5}{2}$ | 0 | 5.92 |

Adding in the full orbital contribution like this is obviously an overestimate. It would appear, then, that there is some orbital contribution, but not as much as implied by Equation 6.11, which applies to the free ions. Here we are concerned with ions in complexes. In a free ion, the electron can circulate around a field along the $z$-axis by going from $d_{xz}$ to $d_{yz}$, which are combinations of orbitals with $m_l = +1$ and $m_l = -1$, or by going from $d_{x^2-y^2}$ to $d_{xy}$, which are combinations of orbitals with $m_l = +2$ and $m_l = -2$. An electron in $d_{z^2}$ has no other orbital to move into with the same value of $m_l$, and does not contribute to the orbital magnetic moment. If the ion is in an octahedral or tetrahedral complex, then $d_{x^2-y^2}$ and $d_{xy}$ no longer have the same energy. The only orbital contribution in these complexes thus comes from $d_{xz}$ and $d_{yz}$. If both orbitals are occupied by electrons of the same spin, then one will have

---

$^\dagger$ $l$ is the orbital momentum quantum number, which defines specific sub-shells. $m_l$ is the magnetic quantum number, which distinguishes the individual orbitals within a sub-shell.

$s$ is the spin quantum number, and $m_s$ is the magnetic spin quantum number.

$m_l = +1$ and one $m_l = -1$, and the magnetic moments will cancel. An orbital contribution to the magnetic moment is therefore only expected if there is a vacancy in $d_{xz}$ or $d_{yz}$. In octahedral complexes, $d_{xz}$ and $d_{yz}$ belong to $t_{2g}$, and in tetrahedral complexes to $t_2$. Thus, we only expect orbital contributions for complexes with 1, 2, 4 and 5 electrons in $t_{2g}$ or $t_2$.

●  Which of the configurations $d^1$ to $d^9$ for weak-field octahedral and tetrahedral complexes are expected to show an orbital contribution to the magnetic moment?

●  Octahedral: $d^1$ $(t_{2g}^1)$, $d^2$ $(t_{2g}^2)$, $d^6$ $(t_{2g}^4e_g^2)$, $d^7$ $(t_{2g}^5e_g^2)$;

   tetrahedral: $d^3$ $(e^2t_2^1)$, $d^4$ $(e^2t_2^2)$, $d^8$ $(e^4t_2^4)$, $d^9$ $(e^4t_2^5)$.

These expectations are borne out for $d^6$ and $d^7$ octahedral complexes, and $d^8$ tetrahedral complexes. There are no data for $d^3$ and $d^4$ tetrahedral complexes.

In general, magnetic measurements are not useful in distinguishing between tetrahedral and weak-field octahedral environments, since both will have the same number of unpaired electrons. There is some correlation between geometry and magnetic moment, however, for complexes of cobalt(II) and nickel(II), due to the differing orbital magnetic moment contributions. For cobalt(II) complexes, which are $d^7$, the octahedral complexes have higher magnetic moments ($4.7$–$5.2\mu_B$) than the tetrahedral complexes ($4.4$–$4.8\mu_B$), but for nickel(II), $d^8$, it is the tetrahedral complexes that have higher orbital contributions, with magnetic moments in the range $3.7$–$4.0\mu_B$, compared with $2.8$–$3.4\mu_B$ for octahedral complexes.

The observed values for tetrahedral nickel(II) and octahedral cobalt(II) complexes show that care must be taken in interpreting the results if large orbital contributions are expected, as the values for these compounds are close to the spin-only values for complexes with one more unpaired electron.

So far, we have only considered tetrahedral and octahedral complexes. In substituted and slightly distorted complexes, although we lose the degeneracy of some of the 3d levels, we generally have the same number of unpaired electrons. For predicting the spin-only magnetic moment, therefore, we can regard such complexes as tetrahedral or octahedral. In some cases, the degeneracy of $d_{xz}$ and $d_{yz}$ is lost, and the magnetic moment will then be closer to the spin-only value than that of true tetrahedral or octahedral complexes.

### QUESTION 6.1

(a)  Explain why high-spin complexes of manganese(II) have magnetic moments close to the spin-only value, $\mu_S$.

(b)  For which strong-field configurations of octahedral complexes would you expect to observe an orbital contribution to the magnetic moment?

### QUESTION 6.2

The compound $K_2[NiF_6]$ is diamagnetic (magnetic moment = 0). What information does this give you about the electronic structure of the complex ion $[NiF_6]^{2-}$?

### QUESTION 6.3

The magnetic moment of $[CoCl_4]^{2-}$ is $4.6\mu_B$ and that of $[Co(H_2O)_6]^{2+}$ is $c.\ 5\mu_B$. What are the electronic structures of the metals in these two complexes? Explain as far as you can the differences in the magnetic moments.

# SUMMARY OF CRYSTAL-FIELD THEORY

7

The main features of crystal-field theory and some of its shortcomings are as follows:

1   Crystal-field theory considers a metal complex as a metal ion surrounded by point negative charges positioned roughly at the coordinating atom of the ligand.

2   These negative charges affect the energies of the orbitals on the metal ion. In particular, the metal 3d orbitals are no longer degenerate, but are split into two or more energy levels. The splitting pattern depends on the symmetry of the complex.

3   For octahedral complexes, the 3d level is split into two, a triply degenerate lower level labelled $t_{2g}$, and a doubly degenerate upper level labelled $e_g$. The energy separation of the two levels is referred to as $\Delta_o$.

4   In tetrahedral complexes, there are again two levels, but this time, the lower level, e is doubly degenerate and the upper level, $t_2$, is triply degenerate. The energy separation of the two levels is referred to as $\Delta_t$, which is roughly $\frac{4}{9}\Delta_o$.

5   For distorted octahedral or square-planar complexes, d-orbital splitting patterns are similarly dictated by the orientation of the d orbitals with respect to the symmetry of the complex.

6   To obtain the electronic configuration of a transition metal in a complex, we have to think of the 3d electrons in the free transition-metal ion going into the appropriate split levels for the particular symmetry of the complex.

7   The magnitude of $\Delta$ depends on the nature of the ligands and the metal ion, and the oxidation state of the metal. If the gap is small, electrons enter the orbitals of the upper level when all the orbitals of the lower level contain electrons of one spin. This is called the *weak-field case*. On the other hand, if the energy gap is large, electrons pair up in the lower level before entering the upper level. This is called the *strong-field case*.

8   Metal orbitals that point towards the ligands shield the ligands from the metal ion more than those that point between the ligands; this affects the ionic radius of the metal and hence the metal–ligand distance.

9   The amount of energy an ion gains by being in a non-spherical environment is called the *crystal-field stabilisation energy* (CFSE). The variation in CFSE across the first transition series explains the double-bowl deviations from a smooth curve for some thermodynamic properties.

10   Selection rules dictate whether a spectroscopic transition is probable or improbable. An allowed electronic transition obeys the Laporte selection rule ($\Delta l = \pm 1$) and the spin selection rule ($\Delta s = 0$).

11   Some low-intensity bands in the electronic spectra of transition-metal complexes can be assigned to transitions of an electron from one of the levels into which the 3d splits to another (for example, $t_{2g} \rightarrow e_g$). These are known as d $\leftrightarrow$ d bands.

12   Magnetic measurements can be used to determine the number of unpaired electrons in a complex. Hence they can distinguish between strong-field and weak-field complexes and, in some cases, between different geometries.

There are, however, a few problems. Firstly, crystal-field theory has no basis for explaining the position of various ligands in the spectrochemical series. We also noted that many of the deepest colours of transition-metal complexes were due to charge-transfer rather than d $\leftrightarrow$ d transitions.

# MOLECULAR ORBITAL THEORY OF TRANSITION-METAL COMPLEXES

8

As you have seen, crystal-field theory can be used to explain many properties of transition-metal complexes, but there are occasions when we need a more accurate theory. For example, the position of ligands in the spectrochemical series was purely empirical; we had no satisfactory explanation for the order. Indeed, we noted in Section 3 that a simple electrostatic model led to a prediction at odds with the empirical order of ligands in the series. Again, we noted that many of the deepest colours found for transition-metal complexes are not due to $d \leftrightarrow d$ transitions, but to charge-transfer transitions.

In crystal-field theory, we regarded the ligands as negative point charges and considered the effect of these on the metal 3d atomic orbitals. In molecular orbital theory we form *molecular* orbitals for the entire complex by combining the metal 3d orbitals with orbitals on the ligands. This approach not only deals with some of the problems noted for crystal-field theory but is also the basis for accurate calculations of properties of transition-metal complexes.

Advances in computer hardware in the last decades of the twentieth century mean that *ab initio* methods (that is, molecular orbital methods that solve the Schrödinger equation without using information from experiments) can be applied to transition-metal complexes. In the Hartree–Fock (HF) approach, electrons are assigned to orbitals, and the energy of each electron in turn is calculated assuming an average distribution of the other electrons. The orbital of each electron is varied until a minimum energy is obtained, and the new orbital is used to produce the average distribution for the next electron. This process continues until no further change is needed to the orbital for any electron. The orbitals thus obtained are used to calculate properties of the complex such as most stable geometry and spectroscopic parameters. Density functional theory (DFT) is based on the realisation that the exact ground-state energy of a molecule can be expressed as a mathematical function of the density known as the *density functional*. The form of the density functional is not completely known, and so various approximations have been developed and different ones chosen to suit the type of calculation required. The electron density is usually calculated by assuming that electrons occupy orbitals as in the Hartree–Fock method, and an energy minimisation is used to obtain the orbitals. The DFT method can make some allowance for the fact that electrons do not simply experience an average charge distribution, but experience local interactions with the other electrons.

These days, it is possible to perform molecular orbital calculations on transition-metal complexes and even organometallic compounds. Using the nature and energy of the orbitals so obtained, we can calculate properties such as the number of unpaired electrons, vibrational frequencies and NMR chemical shifts. For compounds containing transition metals, density functional methods have proved particularly useful. Table 8.1 shows some results of calculations on $TiF_4$, comparing the calculated Ti—F bond distance obtained by minimising the energy from *ab initio* Hartree–Fock (HF) calculations and density functional theory (DFT) calculations, with the experimental value.

**Table 8.1** Predictions of Ti–F distances in TiF$_4$ compared with the empirical value

| Method | Basis set size | Ti–F bond distance/pm |
|---|---|---|
| Hartree–Fock | small | 171.9 |
| Hartree–Fock | large | 174.6 |
| density functional theory | small | 173.1 |
| density functional theory | large | 176.3 |
| experimental | | 175.4 |

⬤ From Table 8.1, what method would you select for calculating the geometry of a transition-metal complex?

⬤ Both DFT and HF give reasonable results provided a large basis set is used.

DFT methods tend to give longer bond distances than HF methods for the corresponding basis set, as in this Table, but both the HF and DFT values differ from the experimental value by about the same amount (smaller and larger, respectively). For calculating other properties, such as bond distances in carbonyl complexes and organometallic compounds, DFT can sometimes do a lot better. Accurate calculated values for $^{59}$Co NMR chemical shifts, for example, were only obtained after the introduction of DFT methods.

We are now going to look at what type of molecular orbitals we find for transition-metal complexes. We start by reminding you how molecular orbitals are formed for small molecules containing only atoms of the main-Group elements.

For these molecules, the orbitals are built up using atomic orbitals on the atoms in the molecule. In order to decide which atomic orbitals to combine, we use the following guidelines:

- atomic orbitals that combine must be of similar energy;
- only atomic orbitals of the same symmetry can combine;
- there must be significant overlap of combining orbitals;
- $n$ atomic orbitals combine to make $n$ molecular orbitals.

For diatomic molecules, for example, the first point means that we usually only consider valence orbitals; the second point means that s orbitals combine with p$_z$, but not p$_x$ or p$_y$; the third point excludes overlap of a core orbital on one atom and a valence orbital on the other; the fourth point means that six molecular orbitals form from the three p orbitals on each atom.

For transition-metal complexes, we need to decide which orbitals on the ligands will combine with the d orbitals on the metal. For simple complexes such as $[MCl_6]^{n-}$, we consider the valence atomic orbitals on the ligand atom that are closest in energy to the metal d orbitals. However, for most complexes the ligands are not single atoms but molecules. The general strategy here is first to take molecular orbitals for the individual ligand molecules, and then consider which of these molecular orbitals will combine with the metal d orbitals. The ligand molecular orbitals that are of the right energy to combine with metal d orbitals are usually the highest-occupied and lowest-unoccupied orbitals. For example, Figure 8.1 shows a partial molecular orbital energy-level diagram for CO and boundary surfaces of the corresponding molecular orbitals.

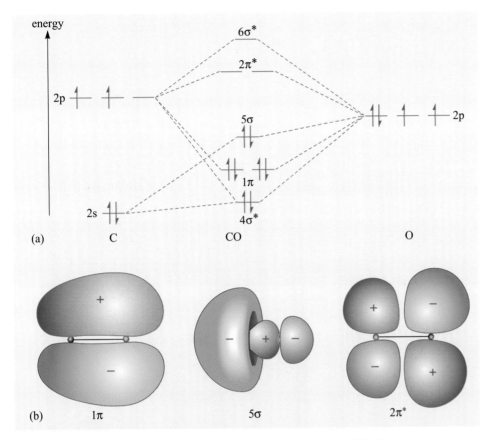

**Figure 8.1** (a) Partial orbital molecular energy-level diagram for CO; (b) boundary surfaces of the 1π, 5σ and 2π* molecular orbitals.

To form orbitals for a transition-metal carbonyl complex (that is, one with CO ligands), we would consider the occupied 5σ and 1π orbitals, and the unoccupied 2π* and 6σ* orbitals formed from the C 2p and O 2p atomic orbitals. We can divide the ligand orbitals into two sets — σ-bonding orbitals, which overlap with the metal orbital to increase electron density along the metal–ligand bond, and π-bonding orbitals, where the electron density is above and below the bond. In Section 9 we start by looking at bonding in octahedral complexes.

# BONDING IN OCTAHEDRAL COMPLEXES

9

## 9.1 σ-bonding

Examples of ligand orbitals which will form σ orbitals with metal d orbitals are s and p orbitals on halide ions, σ orbitals on diatomic ligands such as $OH^-$ and CO, and bonding molecular orbitals formed from H 1s and N or O 2p atomic orbitals in ligands such as $H_2O$ and $NH_3$. We shall represent a generalised **σ-bonding ligand orbital** by a tear-shape as in Figure 9.1. Filled orbitals of this type are usually of lower energy than the metal d orbitals, and so we shall assume this is the case in the subsequent discussion.

For an octahedral complex, we start with a metal ion surrounded by six ligands, each contributing one filled orbital. Thus, we have five metal d orbitals and six ligand orbitals from which to construct molecular orbitals for the complex. We arrange the six ligands to lie on the $x$-, $y$- and $z$-axes. Let us now see how the six ligand orbitals overlap with the metal d orbitals.

Figure 9.2 shows a generalised σ-bonding ligand orbital on the $z$-axis overlapping with the $d_{z^2}$ orbital on the metal.

- Will such a σ-bonding orbital combine with the $d_{z^2}$ orbital?
- Yes. The orbitals have lobes of the same sign overlapping, and so can form a bonding orbital.

The other ligand orbital on the z-axis will also overlap in this manner, and the ligand orbitals along the $x$- and $y$-axes will overlap with the torus. So the metal $d_{z^2}$ orbital can combine with all six ligand orbitals. In other words, there is a combination of the six σ-bonding ligand orbitals that has the same symmetry as the metal $d_{z^2}$ orbital. This combination of six ligand orbitals with the metal $d_{z^2}$ orbital produces two orbitals for the complex — one bonding and one antibonding. Figure 9.3 shows this combination of ligand orbitals, and its interaction with the metal $d_{z^2}$ orbital to form the bonding and antibonding orbitals of the complex.

Figure 9.4 shows the overlap of a ligand orbital on the z-axis with a metal $3d_{x^2-y^2}$ orbital. Here there is no net overlap because the bonding overlap with the positive lobes is cancelled out by the antibonding overlap with the negative lobes.

Ligand orbitals on the $x$- and $y$-axes, however, do have a net overlap with the metal $3d_{x^2-y^2}$ orbital (Figure 9.5). The $d_{x^2-y^2}$ orbital, therefore, can combine with a combination of four ligand orbitals of the same symmetry as itself. Again, one bonding and one antibonding orbital are formed from the $d_{x^2-y^2}$ orbital and this combination of ligand orbitals. Figure 9.5 shows the combination of four ligand orbitals overlapping with the metal $d_{x^2-y^2}$ orbital to form a bonding orbital.

Thus, $d_{z^2}$ and $d_{x^2-y^2}$ orbitals on the metal will overlap with σ-bonding orbitals on the ligands to form one bonding and one antibonding orbital each for the metal complex.

**Figure 9.1** A generalised σ-bonding ligand orbital.

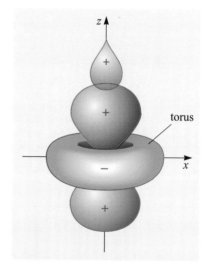

**Figure 9.2** Overlap of a σ-bonding ligand orbital with the $d_{z^2}$ orbital on the metal atom.

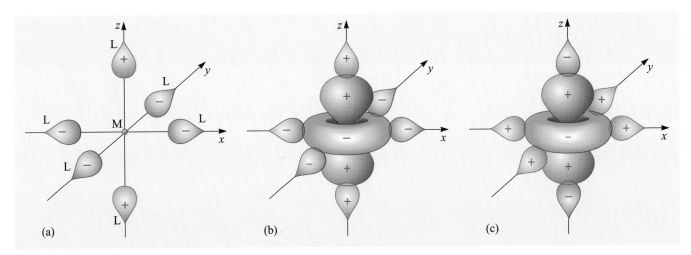

**Figure 9.3** (a) Combination of six σ-bonding ligand orbitals around a metal atom;
(b) the bonding orbital for the complex formed from the ligand orbital combination and the
metal $d_{z^2}$ orbital; (c) the corresponding antibonding orbital for the complex.

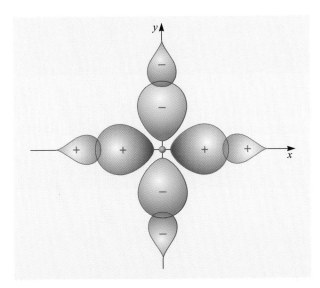

**Figure 9.5** A bonding combination of σ-bonding ligand
orbitals on the *x*- and *y*-axes overlapping with a $3 d_{x^2 - y^2}$ orbital
on a metal atom.

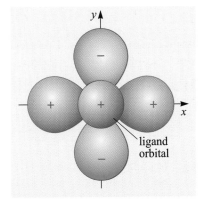

**Figure 9.4** A combination of a
σ-bonding ligand orbital on the
*z*-axis and a metal $3 d_{x^2 - y^2}$ orbital
results in no net overlap.

From two metal d orbitals and six ligand orbitals, we have made two bonding
orbitals and two antibonding orbitals. But from eight orbitals we can make
eight molecular orbitals for the complex. The remaining four orbitals will be
non-bonding combinations of ligand orbitals, unless we can combine these
with the other d orbitals. So we overlap the σ-bonding ligand orbitals with
$d_{xy}$, $d_{yz}$ or $d_{xz}$? Figure 9.6 shows a σ-bonding ligand orbital on the *z*-axis and a
metal $d_{xz}$ orbital.

⬤ Will the $d_{xz}$ metal orbital form a bonding orbital with σ-bonding ligand
orbitals on the *z*-axis?

⬤ No. Overlap with the positive lobe is cancelled by antibonding overlap
with the negative lobe.

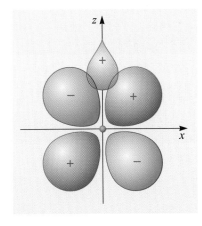

**Figure 9.6** A σ-bonding ligand
orbital on the *z*-axis and a metal $d_{xz}$
orbital.

◯ Will the $d_{xz}$ metal orbital form a bonding orbital with σ-bonding orbitals on the x-axis?

◯ No. As for the σ-bonding ligand orbitals on the z-axis, overlap with the negative lobe will offset that with the positive lobe.

A σ-bonding ligand orbital on the y-axis will overlap with all four lobes but, as for the σ-bonding ligand orbital on the z-axis and the metal $d_{x^2-y^2}$ orbital, the net effect is non-bonding. So the metal $d_{xz}$ orbital does not form any bonding orbitals for the complex with σ-bonding ligand orbitals. The $d_{xy}$ and $d_{yz}$ orbitals behave similarly, so that the three metal orbitals $d_{xy}$, $d_{yz}$ and $d_{xz}$ remain non-bonding in the presence of σ-bonding ligand orbitals. As in crystal-field theory, these three orbitals are degenerate; we label them with the appropriate symmetry label, $t_{2g}$.

Calculations of the energies of the bonding orbitals formed from the metal $d_{z^2}$ and $d_{x^2-y^2}$ orbitals show that the bonding orbitals are also degenerate. These are labelled $e_g$, the symmetry label for $d_{z^2}$ and $d_{x^2-y^2}$ orbitals in an octahedral environment. The antibonding complex orbitals must also be degenerate; they have the same symmetry, so are labelled $e_g$*. Figure 9.7 shows a partial energy-level diagram for σ-bonded octahedral complexes; it comprises the metal d orbitals, the ligand σ-bonding orbitals, and the complex orbitals formed from them. On the left, are the metal d orbitals, on the right, and at lower energy, are the ligand σ-bonding orbitals. The lowest-energy complex orbitals are the $e_g$ bonding orbitals. These will be lower in energy than both the metal and ligand orbitals. As you saw when we were discussing the formation of the bonding orbitals, there are four non-bonding combinations of ligand orbitals. These will lie at the same energy as the ligand orbitals; they have been left out of Figure 9.7 for clarity. The non-bonding metal $d_{xy}$, $d_{yz}$ and $d_{xz}$ orbitals form the $t_{2g}$ orbitals at the same energy level as the metal d orbitals. Finally, above both the metal and ligand orbitals, we have the antibonding $e_g$* orbitals.

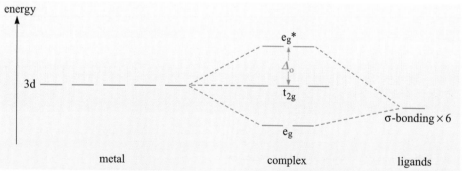

**Figure 9.7** Partial energy-level diagram for a σ-bonded octahedral complex.

The σ-bonding ligand orbitals contain two electrons each, so that we have twelve electrons from these orbitals to put into the complex orbitals. We can assign four of these electrons to the $e_g$ bonding orbitals †. The other eight are assigned to the four non-bonding combinations of ligand orbitals. This leaves the d electrons from the metal ion to fill the $t_{2g}$ and $e_g$* orbitals. These orbitals are analogous to the $t_{2g}$ and $e_g$ atomic orbitals in crystal-field theory. The crystal-field splitting is now replaced by

---

† Strictly speaking, electrons from the ligands and the metal are indistinguishable, but for the purposes of filling in energy-level diagrams we shall notionally indicate the electrons as being derived from the ligands or the metal.

the energy difference between the $t_{2g}$ and $e_g^*$ levels, which is labelled as $\Delta_o$ in Figure 9.7. These levels and the energy difference $\Delta_o$ between them correspond to those obtained by crystal-field theory, and are shown in green in Figure 9.7 and similar figures. To indicate that we are no longer confined to the simple crystal-field model, we shall from now on refer to the *ligand field* so that the energy difference between $t_{2g}$ and $e_g^*$ becomes the **ligand-field splitting energy**.

Using a reference state in which the $t_{2g}$ and $e_g^*$ levels are equally occupied, we can work out the relative stabilities of transition-metal states, and thereby obtain a quantity equal to the crystal-field stabilisation energy, which we shall call the **ligand-field stabilisation energy, LFSE**. For example, a complex of a $d^1$ metal ion will have one electron in $t_{2g}$ and none in $e_g^*$. We can define a reference state as one in which the electron occupies $t_{2g}$ and $e_g^*$ equally, so that it spends $\frac{3}{5}$ of its time in $t_{2g}$ and $\frac{2}{5}$ of its time in $e_g^*$. Since the $e_g^*$ level is $\Delta_o$ higher in energy than the $t_{2g}$, by spending all its time in $t_{2g}$, the electron gains $\frac{2}{5}\Delta_o$. Hence, the complex will be more stable by this amount. If you look back at Table 2.1, you will see that this is equal to the crystal-field stabilisation energy for a $d^1$ octahedral complex.

⦿ What will be the orbital occupancies and ligand-field stabilisation energy for a high-spin octahedral σ-bonded complex of a $d^7$ ion?

⦿ Five of the seven electrons will occupy $t_{2g}$ and two will go into $e_g^*$, giving a configuration $t_{2g}^5 e_g^{*2}$. Each electron in $t_{2g}$ will contribute a ligand-field stabilisation energy of $\frac{2}{5}\Delta_o$, and each electron in $e_g^*$ will contribute $-\frac{3}{5}\Delta_o$. The total ligand-field stabilisation energy will thus be $(5 \times \frac{2}{5}\Delta_o) - (2 \times \frac{3}{5}\Delta_o) = \frac{4}{5}\Delta_o$, in agreement with the value in Table 2.1.

As well as simply enabling us to work out LFSEs, diagrams such as Figure 9.7 give us some insight into what makes a particular ligand strong field or weak field. The size of the ligand-field splitting energy, $\Delta_o$, will depend on the strength of the σ-bonding between metal and ligand; the stronger the bonding the greater will be the energy gap and the stronger the ligand field. To form a strong bond, the ligand must have a filled σ-bonding orbital close in energy to that of the metal d orbitals, and which overlaps well with the d orbitals.

### QUESTION 9.1

Write down the electronic configuration for strong-field and weak-field σ-bonded $d^6$ transition-metal complexes.

σ-bonding by itself, however, is not enough to explain the spectrochemical series; the strongest-field ligands such as CO and $PR_3$ owe their strong metal–ligand bonds to their ability to form π bonds as well as σ bonds. It is to this π-bonding that we now turn. This will enable us to begin to account for the somewhat unexpected ordering of ligands in the spectrochemical series described in Section 3.

# 9.2 π-bonding in strong-field complexes

**π-bonding ligand orbitals** are antisymmetric to rotation about the metal–ligand bond; they form complex orbitals in which the electron density is concentrated above and below the bond (as in the π orbitals of diatomic molecules). Of particular importance for strong-field ligands are empty π-bonding orbitals whose energy is

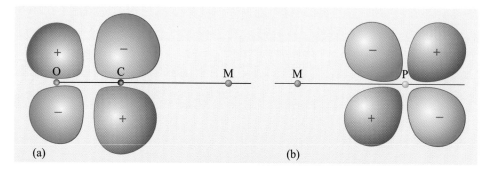

**Figure 9.8** Empty π-bonding orbitals on strong-field ligands: (a) 2pπ* on CO; (b) 3d on phosphorus in phosphines, PR$_3$.

close to that of the metal d orbitals. Examples are the 2pπ* in CO and CN$^-$, and phosphorus 3d orbitals in phosphines. In Figure 9.8, these orbitals are drawn showing their orientation with respect to the M–ligand bond when available for π-bonding.

Let us see which metal d orbitals will overlap with such ligand orbitals. In Section 9.1 we saw that with purely σ-bonding ligands, the d$_{xy}$, d$_{yz}$ and d$_{xz}$ orbitals remained non-bonding. Do they form π bonds? Figure 9.9 shows the overlap of a π-bonding ligand orbital along the x-axis with a metal d$_{xz}$ orbital.

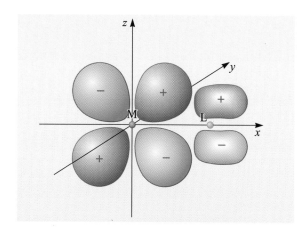

**Figure 9.9** Overlap of a π-bonding ligand orbital on the x-axis with a metal d$_{xz}$ orbital.

○ Will this combination of ligand and metal orbital form a bonding molecular orbital?

● Yes. The positive lobes on each combining pair of orbitals overlap, as do the negative lobes.

This d orbital will interact with two π-bonding ligand orbitals on the x-axis and two on the z-axis as in Figure 9.10 to form a complex π-bonding orbital.

The d$_{xy}$ orbital will form a similar bonding orbital with π-bonding ligand orbitals along the x- and y-axes, and the d$_{yz}$ orbital will form one with π-bonding ligand orbitals on the y- and z-axes. Thus, there will be three bonding complex orbitals formed by the metal d$_{xy}$, d$_{yz}$ and d$_{xz}$ orbitals.

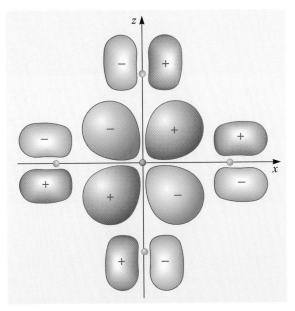

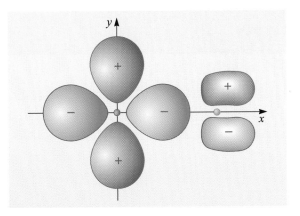

**Figure 9.11** Overlap of a metal $d_{x^2-y^2}$ orbital and a $\pi$-bonding ligand orbital on the $x$-axis.

**Figure 9.10** $\pi$-bonding orbital for a metal complex formed from a metal $d_{xz}$ orbital and $\pi$-bonding ligand orbitals on the $x$- and $z$-axes.

How should these orbitals be labelled?

$t_{2g}$. This label describes the symmetry of the $d_{xy}$, $d_{yz}$ and $d_{xz}$ orbitals of the metal in an octahedral complex, and so any complex orbital formed from them must also have this label.

Thus, we have three degenerate $\pi$-bonding complex orbitals from the metal $d_{xy}$, $d_{yz}$ and $d_{xz}$ orbitals. Will the $d_{z^2}$ and $d_{x^2-y^2}$ metal orbitals form $\pi$ bonds as well as $\sigma$ bonds? Figure 9.11 shows a $\pi$-bonding ligand orbital on the $x$-axis and a metal $d_{x^2-y^2}$ orbital.

The overlap with the positive lobe of the ligand $\pi$-bonding orbital is cancelled out by the overlap with the negative lobe. The net interaction between the $d_{x^2-y^2}$ orbital and $\pi$-bonding ligand orbitals on the other axes is also non-bonding, as is the interaction between the $\pi$-bonding ligands and the $d_{z^2}$ orbital. Thus, as in crystal-field theory, the metal d orbitals in molecular orbital theory are divided into two sets — the $d_{z^2}$ and $d_{x^2-y^2}$ orbitals, which overlap with $\sigma$-bonding ligand orbitals to form $e_g$ complex orbitals, and the $d_{xy}$, $d_{yz}$ and $d_{xz}$ orbitals, which overlap with $\pi$-bonding ligand orbitals to form $t_{2g}$ complex orbitals.

Now we shall build up an energy-level diagram for an octahedral metal complex, showing both $\sigma$- and $\pi$-bonding. The ligand $\sigma$-bonding orbitals are full and of lower energy than the metal d orbitals. The ligand $\pi$-bonding orbitals are empty and of higher energy than the metal d orbitals. So on the left we put the d orbitals for the metal ion. On the right, we have the ligand orbitals, with the $\sigma$-bonding orbitals at lower energy than the metal d orbitals, and the $\pi$-bonding orbitals at higher energy than the metal d orbitals. You saw that there were six $\sigma$-bonding ligand orbitals. How many $\pi$-bonding ligand orbitals are there?

How many $2\pi^*$ orbitals are there in CO?

Two, at right-angles to each other.

Most of the ligand orbitals we are considering, like the $2p\pi^*$ orbitals in CO, are doubly degenerate, so that for the six ligands there are twelve $\pi$-bonding orbitals altogether. A typical energy-level diagram, showing only those complex orbitals involving metal d orbitals is given in Figure 9.12.

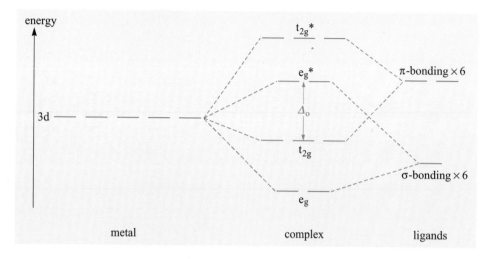

**Figure 9.12** Partial energy-level diagram for an octahedral complex with both $\sigma$- and $\pi$-bonding. All non-bonding ligand orbitals have been omitted for clarity.

For the metal complex, we have a total of 23 orbitals (five d orbitals, six $\sigma$-bonding ligand orbitals and twelve $\pi$-bonding ligand orbitals).

At the lowest energy, we have the two $\sigma$-bonding $e_g$ orbitals. These will be of lower energy than the metal orbitals and the ligand orbitals.

Next will be the four non-bonding $\sigma$-bonding ligand orbitals. (For clarity, we shall leave these out of our diagram, although you will see later that we need to consider them when discussing the spectra of transition-metal complexes.)

Above the $\sigma$-bonding ligand orbitals, but below the metal d orbitals, we then have the three $\pi$-bonding $t_{2g}$ orbitals of the complex.

Above the 3d orbitals, but not necessarily above the $\pi$-bonding ligand orbitals, we have the two antibonding $e_g^*$ orbitals.

Level with the $\pi$-bonding ligand orbitals, we have the nine ligand orbitals of this type, which are not involved in bonding with the metal d orbitals. Again, we shall leave these out of our diagram for now.

Finally, at the highest energy, we have the three antibonding $t_{2g}^*$ orbitals of the complex.

Because the ligand $\pi$-bonding orbitals were empty, we still only have twelve electrons from the ligands. These will fill the two bonding $e_g$ orbitals and the four non-bonding $\sigma$-bonding ligand orbitals as before. The metal d electrons can still be thought of as feeding into the $t_{2g}$ and $e_g^*$ levels. Figure 9.13 shows how the electrons occupy the complex orbitals in Figure 9.12 for the complex $[Fe(CN)_6]^{3-}$, a strong-field octahedral complex of iron(III). One of each type of the two ligand-field orbitals in the complex is shown in Figure 9.14.

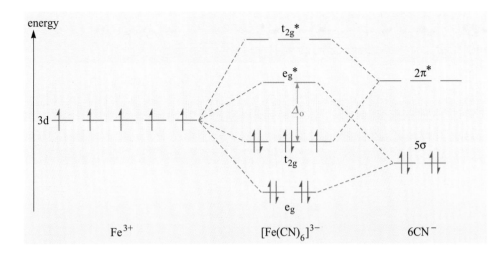

**Figure 9.13** Partial energy-level diagram for $[Fe(CN)_6]^{3-}$. In energy-level diagrams for specific complexes, such as this, we only show for the ligands those orbitals that combine with metal d orbitals. This provides the correct number of electrons to feed into the orbitals of the complex.

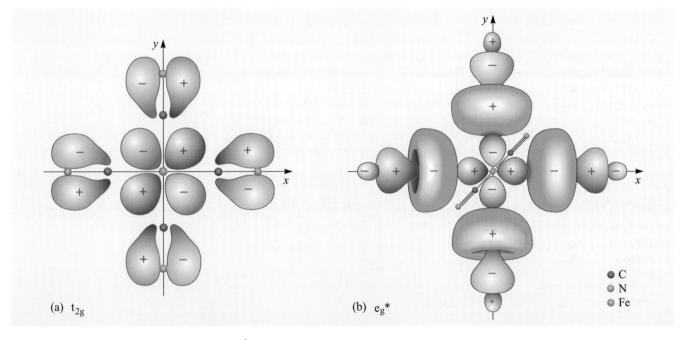

(a) $t_{2g}$      (b) $e_g^*$

● C
● N
● Fe

**Figure 9.14** $t_{2g}$ and $e_g^*$ orbitals of $[Fe(CN)_6]^{3-}$.

● How has including $\pi$-bonding affected the ligand-field splitting, $\Delta_o$, compared with the situation depicted in Figure 9.7?

● $\Delta_o$ has increased because the $t_{2g}$ orbitals are now bonding rather than non-bonding, and are thus lowered in energy.

The strong-field nature of CO and $PR_3$ can thus be explained by their ability to form strong $\pi$ bonds with the $t_{2g}$ metal orbitals, which leads to a lowering of energy of the $t_{2g}$ orbitals of the complex and hence an increase of the energy gap, $\Delta_o$.

QUESTION 9.2

Why do nitrogen ligands, $NR_3$, form weaker-field complexes than the corresponding ligands of phosphorus and arsenic?

# 9.3 π-bonding in weak-field complexes

For some ligands such as halide ions, the effect of π-bonding is to weaken the ligand field. These are ligands where the π-bonding orbitals close in energy to the metal d orbitals are filled.

🔵 What orbitals are appropriate for halide ligands?

🔵 These orbitals are the two $n$p orbitals on each ion not involved in σ-bonding.

How do such orbitals weaken the ligand field? Like the σ-bonding orbitals, the filled π-bonding orbitals are lower in energy than the metal d orbitals. They will overlap with the metal $d_{xy}$, $d_{yz}$ and $d_{xz}$ orbitals to form $t_{2g}$ and $t_{2g}^*$ orbitals.

🔵 Where will the $t_{2g}$ orbitals lie on the energy-level diagram for the complex?

🔵 Below the filled ligand π-bonding orbitals.

A partial energy-level diagram showing only those complex orbitals that involve the metal d orbitals is shown in Figure 9.15. Here we have assumed ligands similar to the halide ions, where the σ- and π-bonding ligand orbitals are at the same energy. For ions such as halide ($X^-$), oxide ($O^{2-}$) and nitride ($N^{3-}$), for example, both the σ-bonding and π-bonding ligand orbitals are $n$p orbitals. For such complexes, the $e_g^*$ level is above the $t_{2g}^*$ level. σ bonds are generally stronger than π bonds, and so where the σ-bonding and π-bonding ligand orbitals are at the same energy, the σ-bonded $e_g$ orbitals will lie below the π-bonded $t_{2g}$ orbitals. Consequently, the antibonding $e_g^*$ orbitals will lie *above* the antibonding $t_{2g}^*$ orbitals.

There are six ligands, each with two π-bonding orbitals. Of these twelve π-bonding orbitals, three overlap with metal d orbitals to form the $t_{2g}$ and $t_{2g}^*$ orbitals. The other nine remain as non-bonding ligand orbitals. In addition, each ligand has one filled σ-bonding orbital, making a total of six filled σ-bonding orbitals. Two of these go to form the $e_g$ and $e_g^*$ complex orbitals, and four are non-bonding. In total, there are 18 ligand orbitals, providing 36 electrons. So let us build up the energy-level diagram.

- The lowest energy level in Figure 9.15 is the $e_g$ bonding level.

- Next is the $t_{2g}$ bonding level. Both this and the $e_g$ bonding level lie below the energies of both the d orbitals and the ligand orbitals.

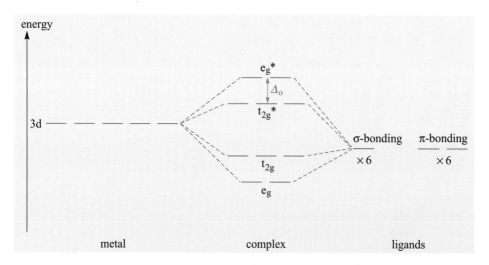

**Figure 9.15** Partial energy-level diagram for an octahedral complex, in which filled σ- and π-bonding ligand (such as halide ions) orbitals interact with the metal d orbitals; non-bonding ligand orbitals have been omitted.

- The nine non-bonding combinations of $\pi$-bonding ligand orbitals and the four non-bonding combinations of $\sigma$-bonding ligand orbitals will have energy levels at about the same energy as the ligand orbital levels. (The levels corresponding to these non-bonding combinations are not shown in Figure 9.15.)

- The next highest energy level for the complex is the $t_{2g}^*$ antibonding level. This is higher in energy than the d orbitals and the ligand orbitals.

- Finally, there is the $e_g^*$ level.

Four electrons from the 36 provided by the ligands fill the $e_g$ orbital of the complex, and six fill the $t_{2g}$ orbital. The remaining 26 electrons fill the 13 non-bonding ligand orbitals, which will be at the same energy as the original ligand orbitals.

⬤ Which orbitals will the electrons from the metal go into?

⬤ The $t_{2g}^*$ and $e_g^*$.

So the electrons from the metal will go to fill the $t_{2g}^*$ and $e_g^*$ orbitals. The ligand-field splitting $\Delta_o$ is now between the $t_{2g}^*$ and $e_g^*$ levels. The effect of the filled $\pi$-bonding orbitals is to replace the non-bonding $t_{2g}$ level of the $\sigma$-bonded complex by an antibonding $t_{2g}^*$ level. Because the antibonding level is higher in energy, the gap between the orbitals is reduced and hence the ligand-field splitting is less than in Figure 9.7. Figure 9.16 shows how the electrons occupy the levels in Figure 9.15 for the weak-field iron(III) complex $[FeF_6]^{3-}$.

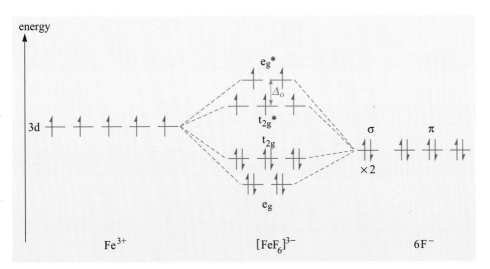

**Figure 9.16** Partial orbital energy-level diagram for $[FeF_6]^{3-}$.

⬤ What will be the electronic configuration for a weak-field $d^4$ complex, in which ligands have filled $\pi$-bonding orbitals?

⬤ $t_{2g}^{*3}e_g^{*1}$.

QUESTION 9.3

Write down the electronic configurations for a $d^5$ complex with: (a) $\sigma$-bonding only (weak field and strong field); (b) $\sigma$-bonding and $\pi$-bonding through empty ligand orbitals (strong field); (c) $\sigma$-bonding and $\pi$-bonding through filled ligand orbitals (weak field).

We can now summarise the factors that affect the strength of the ligand field for octahedral complexes:

(i) Strong overlap between metal d orbitals and filled ligand σ-bonding orbitals close in energy leads to a strong field.

(ii) Overlap between metal d orbitals and empty π-bonding ligand orbitals of slightly higher energy strengthens the ligand field.

(iii) Overlap between metal d orbitals and filled π-bonding ligand orbitals lower in energy than the metal orbitals weakens the ligand field.

Thus, we would expect to find that a ligand with filled σ-bonding orbitals and empty π-bonding orbitals close in energy to the metal d orbitals would be a very strong-field ligand, and such is the case. Examples of ligands fulfilling these criteria are $CO$, $CN^-$ and $PR_3$. The fluoride ion, on the other hand, with a filled π-bonding orbital of the right energy, but no available empty π-bonding orbitals, is a very weak-field ligand.

Many ligands have both filled and empty π-bonding orbitals available; chloride ions, for example, have both filled 3p and empty 3d orbitals available. In this case, the resultant ligand field depends on which π-bonding orbitals give the better overlap with the metal d orbitals. For $Cl^-$, the 3d orbitals do not play a major role, and so $Cl^-$ is a weak-field ligand.

It was noted earlier that $O^{2-}$ was a weaker-field ligand than $H_2O$, although on the purely electrostatic crystal-field model we might have expected the order to be reversed. Let us see if we can explain the relative field strengths of these ligands using molecular orbital theory.

● Which $O^{2-}$ orbitals will combine with metal d orbitals to form orbitals for the metal complex?

● The filled 2p orbitals.

One of these 2p orbitals on each ligand lies along the metal–ligand bond, and two combinations of these will form $e_g$ and $e_g{}^*$ orbitals with the metal $d_{z^2}$ and $d_{x^2-y^2}$ orbitals. The other two 2p orbitals on each ligand will lie above and below the metal–ligand bond. Three combinations of these will form $t_{2g}$ and $t_{2g}{}^*$ orbitals with the metal $d_{xy}$, $d_{yz}$ and $d_{xz}$ orbitals. The energy-level diagram when the ligands are $O^{2-}$ is thus the same as Figure 9.15.

The orbital energy-level diagram for the water molecule, $H_2O$, is given in Figure 9.17; the symmetry labels of the water molecular orbitals are indicated (see Section 10). Figure 9.18 shows the orientation of a water molecule bonded to a metal atom in a complex such as $[Fe(H_2O)_6]^{2+}$.

The $a_1$ (higher-energy) bonding orbital (Figure 9.19) will overlap with $d_{z^2}$ and $d_{x^2-y^2}$ orbitals on the metal. Hence, combinations of this orbital on the water ligands with the metal d orbitals will produce the $e_g$ and $e_g{}^*$ orbitals of the complex (as in Figure 9.15). As well as this filled σ-bonding orbital, each water ligand has two filled π-bonding orbitals. One of these is the non-bonding orbital ($b_2$), which is an O 2p orbital. This will interact with metal orbitals in a similar way to the filled 2p π-bonding orbitals in $O^{2-}$. The other π-bonding orbital is the lower-energy bonding orbital ($b_1$) in Figure 9.17, which will overlap with $d_{xy}$, $d_{yz}$ and $d_{xz}$ orbitals on the metal. This is shown in Figure 9.20. Note that this orbital is a σ orbital in $H_2O$ itself, but because it has two lobes of opposite signs, it acts as a π-bonding orbital when attached to a metal.

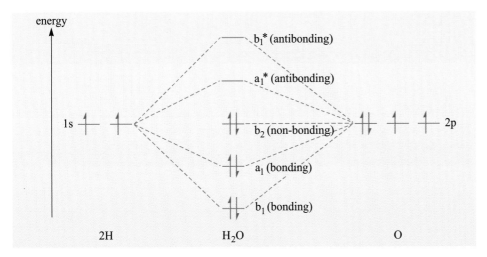

Figure 9.17 Partial energy-level diagram for $H_2O$, showing symmetry labels for the molecular orbitals.

The $b_1$ orbital in Figure 9.20 has less electron density towards the metal than would an O 2p orbital. This orbital will form a weaker $\pi$ bond with the metal than the non-bonding orbital or the O 2p orbitals of $O^{2-}$. If the $\pi$-bonding is weaker, the $t_{2g}*$ orbital will be less antibonding and thus lower in energy. This will increase the energy gap between the $t_{2g}*$ orbital and the $e_g*$ orbital, thus making the ligand stronger field.

The stronger ligand field of $H_2O$ compared with $O^{2-}$ can thus be attributed to its forming weaker $\pi$ bonds to the metal.

## QUESTION 9.4

The 2p orbitals on nitrogen in the ammonia molecule form three bonding molecular orbitals, which are fully occupied (Figure 9.21). One of these is a $\sigma$-bonding orbital ($a_1$), which forms a strong metal–ligand bond. The other two are $\pi$-bonding orbitals (e), both of which contain H 1s as well as N 2p contributions, and the electron density in them is concentrated more towards the hydrogen atoms (that is, away from the metal) than it would be in an N 2p orbital. Explain why $NH_3$ is a stronger-field ligand than nitride, $N^{3-}$.

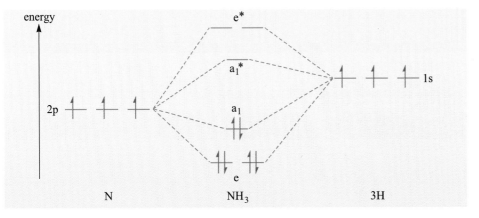

Figure 9.21 Partial energy-level diagram for ammonia, $NH_3$.

Figure 9.18 Orientation of a water molecule bonded to an iron atom in an iron aquo complex.

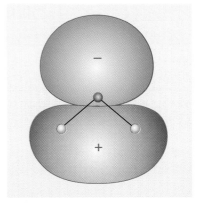

Figure 9.19 Higher-energy bonding orbital ($a_1$) of $H_2O$.

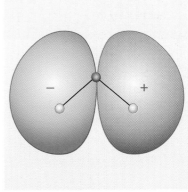

Figure 9.20 Lower-energy bonding orbital ($b_1$) of $H_2O$.

# 9.4 Summary of Section 9

In the molecular orbital theory of metal complexes, bonding in the complex arises from the overlap of metal d orbitals with orbitals on the ligands of compatible symmetry and energy to form orbitals for the complex.

1   Where the ligand is a molecule, we first form molecular orbitals for the ligand, and then use these to overlap with the metal d orbitals.

2   Ligand orbitals with electron density along the metal–ligand bond ($\sigma$-bonding orbitals) form $e_g$ and $e_g{}^*$ orbitals with the metal $d_{z^2}$ and $d_{x^2-y^2}$ orbitals. The $\sigma$-bonding ligand orbitals are filled, and four electrons from them can be thought of as occupying the $e_g$ orbitals. The $e_g{}^*$ orbitals thus play the role of the $e_g$ orbitals in crystal-field theory.

3   If there is only $\sigma$-bonding of the metal to the ligand, then the metal $d_{xy}$, $d_{yz}$ and $d_{xz}$ orbitals remain non-bonding, but are labelled $t_{2g}$ because they are in an octahedral complex. These orbitals play the role of the $t_{2g}$ orbitals in crystal-field theory.

4   Empty ligand orbitals in which the electron density lies above and below the metal–ligand bond ($\pi$-bonding orbitals) interact with the metal $d_{xy}$, $d_{yz}$ and $d_{xz}$ orbitals to form $t_{2g}$ and $t_{2g}{}^*$ orbitals in the complex. Because these ligand orbitals are empty, the $t_{2g}$ bonding orbital acts as the $t_{2g}$ orbital in crystal-field theory. Being a bonding orbital, the $t_{2g}$ orbitals lie lower in energy than the metal d orbitals and the gap between these orbitals and the $e_g{}^*$ orbitals is increased, giving a larger ligand-field splitting.

5   Filled $\pi$-bonding orbitals on the ligand also form $t_{2g}$ and $t_{2g}{}^*$ orbitals in the complex, but there are now electrons from the ligand available to fill the $t_{2g}$ orbitals, so that the $t_{2g}{}^*$ orbitals are equivalent to the $t_{2g}$ orbitals in crystal-field theory. The ligand-field splitting is reduced.

6   The order of ligands in the spectrochemical series can be rationalised in terms of strong $\sigma$-bonding and the availability of empty $\pi$-bonding ligand orbitals, which increase the ligand field, and the availability of filled $\pi$-bonding orbitals, which decrease the ligand field.

# BONDING IN COMPLEXES OF $D_{4h}$ SYMMETRY

# 10

We now go on to look at complexes having other symmetries. In this Section, we consider complexes belonging to the symmetry point group $\mathbf{D_{4h}}$, and in Section 11 we consider tetrahedral complexes.

Before we discuss what types of complex belong to $\mathbf{D_{4h}}$, and what their orbital energy-level diagrams look like, we discuss some symmetry concepts.

## 10.1 Symmetry elements and symmetry point groups [†]

*Symmetry elements* are points, axes or planes, which we can use to describe how symmetrical a molecule or other object is. The action associated with a symmetry element is called a *symmetry operation*. If a complex contains a particular symmetry element, the action of the associated symmetry operation leaves the molecule looking unchanged. Simple symmetry elements are:

- *centre of symmetry*, given the symbol i. If a molecule has a centre of symmetry, inversion through the centre of symmetry leaves the complex looking exactly the same. The operation of inversion is given the symbol $\hat{\imath}$ .

- *plane of symmetry*, $\sigma$. If a molecule has a plane of symmetry, reflection through the plane of symmetry leaves the complex looking the same. The operation of reflection is given the symbol $\hat{\sigma}$ .

- *n-fold rotation axis*, $C_n$. A complex containing an $n$-fold rotation axis is turned into an identical-looking complex when rotated through $1/n$ of a revolution about the axis. The operation of rotation is given the symbol $\hat{C}_n$ . The axis of highest order (largest $n$) in a complex is known as the *principal axis*.

If a complex has at least one rotation axis and one plane of symmetry, the plane(s) of symmetry can be described as horizontal or vertical according to how they are disposed relative to the principal axis. Vertical planes of symmetry, labelled $\sigma_v$, contain the principal axis; horizontal planes, labelled $\sigma_h$, are at right-angles to it. Some complexes with more than one rotation axis have planes labelled $\sigma_d$ (d for dihedral). Like $\sigma_v$ planes, these contain the principal axis. However, $\sigma_d$ planes neither contain, nor are at right-angles to, other axes.

In Figure 10.1, some of the symmetry elements of a square-planar molecule are illustrated. $L_1$–$L_4$ are identical ligands, but have been numbered to help you see the effects of the various symmetry operations. Square-planar molecules belong to the symmetry point group $\mathbf{D_{4h}}$. Complexes belonging to $\mathbf{D_{4h}}$ have one set of vertical planes containing a $C_2$ axis and a set (as in Figure 10.1d), which lies between the $C_2$ axes. The latter set is labelled $\sigma_d$.

---

[†] A flow chart for determining the symmetry point group of an object is provided as an Appendix on p. 84.

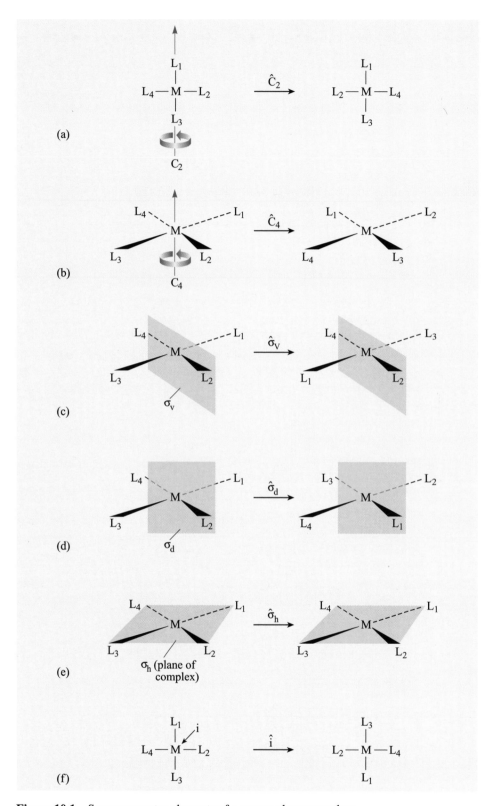

**Figure 10.1**    Some symmetry elements of a square-planar complex:
(a) a $C_2$ axis and its related symmetry operation $\hat{C}_2$; (b) the $C_4$ axis and its related
symmetry operation $\hat{C}_4$; (c) a $\sigma_v$ plane and its related symmetry operation $\hat{\sigma}_v$;
(d) a $\sigma_d$ plane and its related symmetry operation $\hat{\sigma}_d$; (e) the $\sigma_h$ plane and its related
symmetry operation $\hat{\sigma}_h$; (f) the centre of symmetry i and its related symmetry operation $\hat{i}$.

Complexes with the same set of symmetry elements are said to belong to the same *symmetry point group*. Examples of symmetry point groups and their elements are:

**C$_i$**: i

**C$_s$**: σ

**C$_n$**: C$_n$

**C$_{nv}$**: C$_n$ + $n$σ$_v$

**C$_{nh}$**: C$_n$ + σ$_h$ (+ i if $n$ is even)

**D$_{nh}$**: C$_n$ + $n$C$_2$ + σ$_h$ + $n$σ$_v$ (+ i if $n$ is even)

There is another symmetry element and operation that is relevant here, because complexes in groups **T$_d$** and **O$_h$** — that is, perfectly tetrahedral and octahedral complexes — possess such symmetry elements. The element is called an **improper rotation axis**, and is given the symbol S$_n$. Unlike the operations you have met so far, the operation associated with this symmetry element involves two steps. Firstly, we rotate the complex by 1/$n$ of a revolution about the axis, and then reflect the complex through a plane at right-angles to the axis. If the complex contains an S$_n$ axis, the result is an identical-looking complex. In Figure 10.2, this is illustrated for the S$_6$ axis of an octahedral complex. Here we have tipped the complex over, so that instead of four ligands in one plane with the other two above and below the plane, the ligands are arranged to form two triangles. The two triangles are staggered so that the corners of the top one are above a side of the bottom one. You may find this easier to see if you make a model of an octahedral complex and turn it round until it is in the position depicted in Figure 10.2a (left).

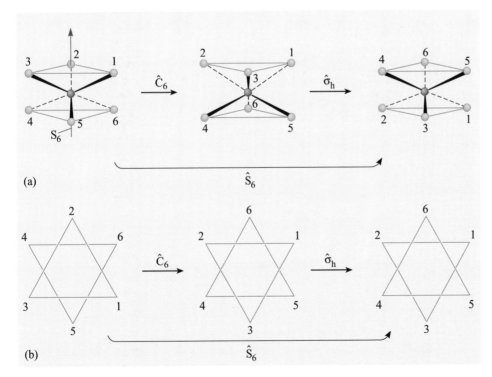

**Figure 10.2** Improper rotation about an S$_6$ axis for an octahedral complex:
(a) side view; (b) plan view.

To completely specify any symmetry point group, we also need to include another operation, the identity operation. This operation is the act of doing nothing at all. It is rather like the mathematical operation of multiplying by 1. The associated symmetry element is the identity element, symbol I. All complexes, whatever their symmetry point group, contain the identity element. Complexes with no axes, planes or centre of symmetry, whose only symmetry element is the identity element, belong to the symmetry point group $\mathbf{C}_1$.

The complete set of symmetry elements of the symmetry point groups $\mathbf{T}_d$ and $\mathbf{O}_h$ are:

$\mathbf{T}_d$: I + $3S_4$ + $4C_3$ + $6\sigma_d$

$\mathbf{O}_h$: I + $4S_6$ + $3S_4$ + $3C_4$ + $6C_2$ + $6\sigma_d$ + $3\sigma_h$ + i

For octahedral and tetrahedral complexes, it is conventional to label the planes containing the principal $S_n$ axis, but between two rotation axes, as *dihedral*.

With the addition of improper rotation axes and dihedral planes, we can now introduce additional point groups. The symmetry point groups $\mathbf{S}_n$ contain only the identity element and an improper rotation axis. Complexes belonging to such groups are rare. The symmetry point groups $\mathbf{D}_{nd}$ contain a $C_n$ axis, $nC_2$ axes, an $S_{2n}$ axis, $n\sigma_d$ planes lying between the $C_2$ axes and, if $n$ is odd, a centre of symmetry. The well-studied organometallic complex ferrocene, in the configuration with the rings staggered (Structure **10.1**), belongs to $\mathbf{D}_{5d}$.

The relevance of symmetry point groups to the study of transition-metal complexes is twofold. Firstly, the pattern of splitting of the metal d energy levels is determined by the symmetry point group to which the complex belongs. You have seen that, for all octahedral complexes and metal ions in octahedral environments in crystals, the metal d levels split into one set of $t_{2g}$ symmetry and one of $e_g$ symmetry. As you saw in Section 2.1, the 't' indicates that the level is triply degenerate, and the 'e' indicates a doubly degenerate level. 'g' and 'u' stand for 'gerade' (German for *even*) and 'ungerade' (German for *odd*), and refer to the behaviour of the orbital on inversion through the centre of symmetry. In tetrahedral symmetry (symmetry point group $\mathbf{T}_d$), the d orbitals are labelled e and $t_2$ because this group does not have a centre of symmetry, so that the subscripts g or u cannot be used. A different splitting pattern to either of these is found for complexes belonging to $\mathbf{D}_{4h}$ with the orbitals labelled differently. The labels, however, are the same for all complexes belonging to $\mathbf{D}_{4h}$, whether they are distorted octahedral or square planar in shape. Complexes belonging to other symmetry point groups, for example $\mathbf{D}_{\infty h}$ (symmetric linear complexes) or $\mathbf{C}_{2v}$ (which includes tetrahedral complexes with two ligands of one type and two of another, $MX_2Y_2$), will have the d orbitals labelled according to their particular symmetry point group.

The second point will come up later when you will see how symmetry determines which spectroscopic transitions are allowed.

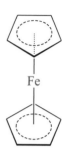

**10.1**

QUESTION 10.1

Using the flow chart in the Appendix, determine the symmetry point group of (a) *trans*-$[FeCl_2Br_4]^{4-}$ and (b) *cis*-$[FeCl_2Br_4]^{4-}$.

# 10.2 Bonding in distorted octahedral complexes

## 10.2.1 Weak-field complexes

As you saw in Section 4, octahedral halide complexes of copper(II) and chromium(II) tend to distort (Jahn–Teller theorem). The octahedra of halogen ions around these metal ions in the crystalline halides usually distorts so that two *trans* halide ligands are further away from the metal than the other four. This is the distortion we consider here, with the two more distant ligands along the $z$-axis. Let us see how this affects the overlap of ligand and metal orbitals.

⬤ With which d orbital(s) on the metal will the σ-bonding orbitals on the $z$-axis overlap?

⬤ The metal $d_{z^2}$ orbital.

⬤ With which d orbital(s) on the metal will the π-bonding orbitals on the $z$-axis overlap?

⬤ The $d_{yz}$ and $d_{xz}$ orbitals.

These d orbitals will be less strongly bonded, since their increased ligand distance from the metal allows less overlap. The remaining d orbitals, which overlap with ligand orbitals from the closer ligands, will be more strongly bonded. Thus, the σ-bonded $e_g$ orbitals in the octahedron will split into two levels, the $d_{z^2}$ (now labelled $a_{1g}$) being higher in energy than the $d_{x^2-y^2}$ ($b_{1g}$). The $e_g{}^*$ orbital levels (which correspond to $e_g$ in crystal-field theory) will also be split, with the $d_{z^2}$ ($a_{1g}{}^*$) orbital *lower* in energy. This splitting is shown in Figure 10.3a.

The π-bonded $t_{2g}$ orbitals will also split into two levels, with the $d_{xy}$ lower in energy ($b_{2g}$) and the $d_{xz}$ and $d_{yz}$ orbitals ($e_g$) higher in energy. Of the corresponding antibonding orbitals, $t_{2g}{}^*$, the one formed from the $d_{xy}$ metal orbital ($b_{2g}{}^*$) will be higher in energy than that formed from the $d_{xz}$ and $d_{yz}$ orbitals ($e_g{}^*$). This splitting is shown in Figure 10.3b. As indicated in Figure 10.3, in $D_{4h}$ symmetry the orbitals formed from the metal $d_{z^2}$ orbital are labelled $a_{1g}$ (and $a_{1g}{}^*$), those formed from the $d_{x^2-y^2}$ orbital are labelled $b_{1g}$ (and $b_{1g}{}^*$), those from $d_{xy}$ are labelled $b_{2g}$ (and $b_{2g}{}^*$), and the degenerate orbitals formed from the $d_{xz}$ and $d_{yz}$ orbitals are labelled $e_g$ (and $e_g{}^*$).

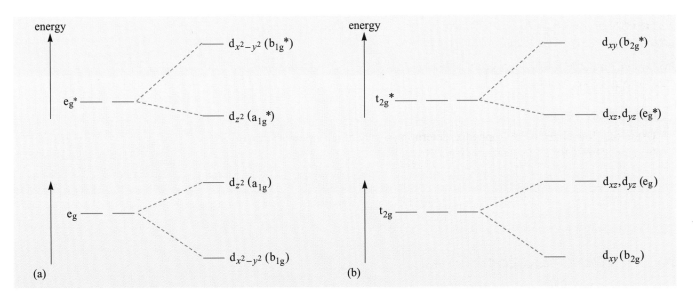

**Figure 10.3**   (a) Energy-level diagrams for the splitting of $e_g$ and $e_g{}^*$ levels in distorted octahedral complexes; (b) energy-level diagrams for the splitting of $t_{2g}$ and $t_{2g}{}^*$ levels in distorted octahedral complexes.

For a complex with filled $\sigma$- and $\pi$-bonding orbitals, the energy-level diagram is illustrated in Figure 10.4. The $b_{1g}$ ($d_{x^2-y^2}$) bonding orbital has the lowest energy. Next comes the other $\sigma$-bonding orbital, the $a_{1g}$ ($d_{z^2}$).

Then there are the $\pi$-bonding orbitals, the $b_{2g}$ ($d_{xy}$) and the $e_g$ ($d_{yz}$ and $d_{zx}$). Electrons from the filled ligand orbitals can be allocated to all these levels, which will thus be filled.

Metal d electrons can then be assigned to the antibonding orbitals. Because the highest-energy (weakest bonding) bonding orbital was $e_g$, the lowest-energy antibonding orbital is $e_g^*$.

The next highest in energy is the $b_{2g}^*$. Then there are the antibonding orbitals from the octahedral $e_g^*$, the $a_{1g}^*$ and, highest of all, the $b_{1g}^*$ orbital.

Complexes with filled $\sigma$- and $\pi$-bonding ligand orbitals are weak field, so that electrons are fed successively into the antibonding levels with parallel spins until all are half-filled. Thus, a $d^3$ complex, for example, will have electronic configuration $e_g^{*2}b_{2g}^{*1}$, and a $d^8$ complex will have electronic configuration $e_g^{*4}b_{2g}^{*2}a_{1g}^{*1}b_{1g}^{*1}$.

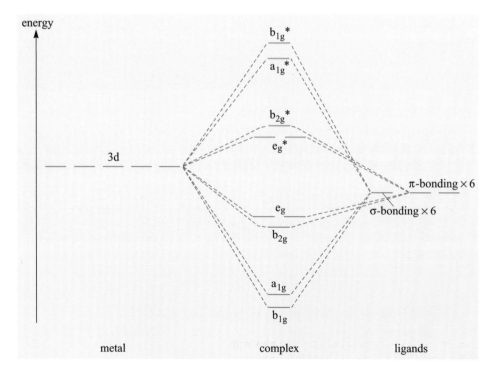

**Figure 10.4** Partial orbital energy-level diagram for a distorted octahedral complex with filled $\sigma$- and $\pi$-bonding ligand orbitals in which two *trans* ligands are further away from the metal than the other four, or two *trans* ligands are weaker-field ligands than the other four. Only orbitals of the complex that involve a contribution from metal d orbitals are shown.

Figure 10.4 would be appropriate for the $Cu^{2+}$ ions in $CuCl_2$, $CuBr_2$ and $CsCuCl_3$, where the metal ion is surrounded by a distorted octahedron of halide ions. The same diagram also applies to complexes *trans*-$MA_4B_2^{n\pm}$, where A and B are weak-field ligands and B is a weaker-field ligand than A.

We now go on to consider similar distortions in strong-field complexes.

QUESTION 10.2

What is the electronic configuration of iron in *trans*-[$FeCl_2(H_2O)_4$]?

## 10.2.2 Strong-field complexes

For complexes with filled σ-bonding ligand orbitals and empty π-bonding ligand orbitals, metal d electrons in an octahedral complex were assigned to the t$_{2g}$ and e$_g$* orbitals. For a small distortion, each of these levels will be split as in Figure 10.3. The energy gap arising from this splitting will be much smaller than the gap between the t$_{2g}$ and e$_g$* orbitals. In the energy-level diagram for such a complex:

- the lowest-energy orbitals are still the b$_{1g}$ and a$_{1g}$, and there are sufficient electrons available from the ligands to fill these;

- the next orbitals are the empty b$_{2g}$ (d$_{xy}$) and e$_g$ (d$_{yz}$ and d$_{xz}$);

- above these are the empty a$_{1g}$* (d$_{z^2}$) and b$_{1g}$* (d$_{x^2-y^2}$) orbitals;

- finally, at the top, are the e$_g$* and b$_{2g}$* orbitals.

d electrons from the metal are assigned to b$_{2g}$, e$_g$, a$_{1g}$* and b$_{1g}$* (Figure 10.5). For a small distortion, the splitting between b$_{2g}$ and e$_g$ will not be large, so that electrons will first occupy these with parallel spins. In strong-field complexes, the gap between e$_g$ and a$_{1g}$* will be large, so that electrons will then pair up in b$_{2g}$ and e$_g$, rather than enter a$_{1g}$* and b$_{1g}$*.

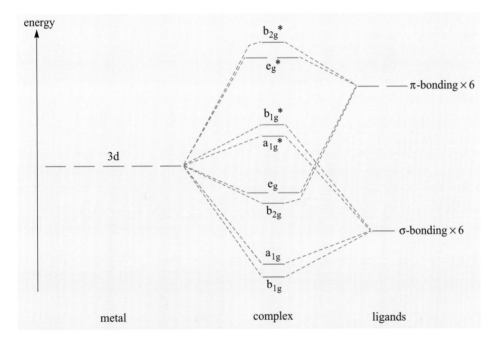

**Figure 10.5** Partial orbital energy-level diagram for a distorted octahedral complex with filled σ-bonding and empty π-bonding ligand orbitals, and two *trans* ligands further away from the metal than the other four, or two *trans* ligands weaker field than the other four.

○ *trans*-[CoF$_2$en$_2$]$^+$ is a strong-field complex. What is the electronic configuration of this complex?

○ F$^-$ is a weaker-field ligand than en, so the orbital energy-level diagram for this complex with the F$^-$ ligands along the z-axis will resemble Figure 10.5. The complex contains cobalt in oxidation state +3 (d$^6$). The six d electrons will fill b$_{2g}$ and e$_g$, giving an electronic configuration b$_{2g}^2$e$_g^4$.

# 10.3 Bonding in square-planar complexes

If we take a distorted strong-field octahedral complex as in Section 10.2, and increase the distortion by moving the ligands on the $z$-axis further away, then the overlap of $d_{z^2}$ with $\sigma$-bonding ligand orbitals, and of $d_{xz}$ and $d_{yz}$ with $\pi$-bonding ligand orbitals will be reduced. Consequently, the $a_{1g}$ and $e_g$ orbitals (Figure 10.5) will be less bonding. For $\sigma$-bonding, the ligand orbitals are lower in energy than the metal d orbitals, and so the $a_{1g}$ orbital will be closer in energy to the $\sigma$-bonding ligand orbitals, and the $a_{1g}^*$ orbital will be closer in energy to the metal d orbitals. The empty $\pi$-bonding ligand orbitals are higher in energy than the metal d orbitals, and so the $e_g$ orbitals will be closer in energy to the metal d orbitals, and the $e_g^*$ orbital will be closer to the empty $\pi$-bonding ligand orbitals. The energy gap between $b_{2g}$ and $e_g$, and that between $a_{1g}^*$ and $b_{1g}^*$ increases.

If the two ligands on the $z$-axis are removed completely, leaving us with a square-planar complex, then the gap between $b_{2g}$ and $e_g$, and between $a_{1g}^*$ and $b_{1g}^*$, increases even more. The gap between $a_{1g}^*$ and $b_{1g}^*$ is often further increased by interaction of the metal 4s orbital (which has $a_{1g}$ symmetry) with the $a_{1g}^*$ orbital of the complex. Because the 4s level is at higher energy than the 3d for metals in oxidation states +2 and +3, this interaction decreases the energy of the $a_{1g}^*$ orbital; it may even drop below the $b_{2g}$ and $e_g$ levels. (Note, however, that the metal 4s–complex $a_{1g}^*$ interaction does not appreciably affect the bonding $a_{1g}$ orbital, as this is much lower in energy than the metal 4s orbital.) The $b_{1g}^*$ orbital, on the other hand, becomes very high in energy relative to the other ligand-field orbitals (shown in green in Figure 10.6).

As with distorted octahedral complexes, there are sufficient electrons from the ligands to fill $a_{1g}$ and $b_{1g}$. The metal d electrons are assigned to $b_{2g}$, $e_g$, $a_{1g}^*$ and $b_{1g}^*$. The order of the three lower levels in this group ($b_{2g}$, $e_g$ and $a_{1g}^*$) varies from complex to complex; if the ligand has suitable filled $\pi$ orbitals, for example, then the $e_g$ level may lie below the $b_{2g}$ level. Thus, the order of levels often differs from that predicted by crystal-field theory. In a typical square-planar nickel(II) complex like $[Ni(CN_4)]^{2-}$, the levels lie in the order $b_{2g}$, $e_g$, $a_{1g}^*$, $b_{1g}^*$ with increasing energy, as shown in Figure 10.6.

The ion $[Ni(PMe_3)_4]^{2+}$ has bands in its electronic spectrum at $16\,600$ cm$^{-1}$ ($a_{1g}^* \leftrightarrow b_{1g}^*$), $20\,400$ cm$^{-1}$ ($e_g \leftrightarrow b_{1g}^*$) and $25\,600$ cm$^{-1}$ ($b_{2g} \leftrightarrow b_{1g}^*$). To a first approximation, the energy difference between $b_{2g}$ and $e_g$ is thus $5\,200$ cm$^{-1}$, between $e_g$ and $a_{1g}^*$ $3\,800$ cm$^{-1}$, but $16\,600$ cm$^{-1}$ between $a_{1g}^*$ and $b_{1g}^*$, illustrating the point made above that the $b_{1g}^*$ is very high in energy relative to the other orbitals.

QUESTION 10.3

The electronic spectrum of the square-planar halide complexes of platinum and palladium, such as $[PdBr_4]^{2-}$, indicate that the order of the ligand-field energy levels in these complexes is $a_{1g}$, $e_g$, $b_{2g}$, $b_{1g}$ (note that the absence of asterisks is purely because the experiment only determined the symmetry; it does not imply that all these levels are bonding). Suggest one reason why this order is different from that of square-planar nickel(II) complexes with strong-field ligands.

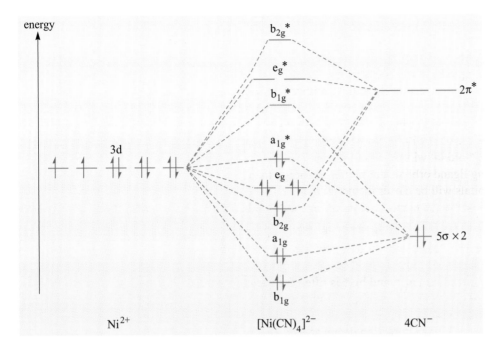

**Figure 10.6** Partial orbital energy-level diagram for $[Ni(CN_4)]^{2-}$, a typical square-planar nickel(II) complex.

## 10.4 Summary of Section 10

1   The symmetry of a complex can be described in terms of its possession of particular symmetry elements.

2   Symmetry elements relevant to molecules (including complexes) are the identity element I, centre of symmetry i, plane of symmetry $\sigma$, $n$-fold rotation axis $C_n$, and $n$-fold improper rotation axis $S_n$.

3   If a molecule contains a particular symmetry element, then the operation associated with that symmetry element leaves the molecule looking unchanged.

4   Complexes with the same set of symmetry elements belong to the same symmetry point group.

5   An important symmetry point group for transition-metal complexes is $\mathbf{D}_{4h}$. Complexes belonging to $\mathbf{D}_{4h}$ include square-planar complexes, distorted octahedral complexes with two *trans* ligand–metal bonds longer than the other four, and substituted octahedral complexes with two *trans* ligands different from the other four, $MX_4Y_2$.

6   In $\mathbf{D}_{4h}$ complexes, what was originally the $e_g$ level in an octahedral complex splits into two, labelled $b_{1g}$ and $a_{1g}$. The $t_{2g}$ level splits into a level labelled $b_{2g}$ and a doubly degenerate level, labelled $e_g$.

7   In weak-field complexes, the ligand-field levels are the $e_g^*$, $b_{2g}^*$, $a_{1g}^*$ and $b_{1g}^*$. The order of these levels can vary. There is no single ligand-field splitting corresponding to $\Delta_o$ or $\Delta_t$ for $\mathbf{D}_{4h}$ complexes.

8   In strong-field complexes, the ligand-field orbitals are the $b_{2g}$, $e_g$, $a_{1g}^*$ and $b_{1g}^*$. Again, the order of the levels may vary.

9   Interaction of the metal 4s orbital with the ligand orbitals leads to a lowering of the energy of the $a_{1g}^*$ level (but not the $a_{1g}$). This is particularly important for square-planar complexes, where the $b_{1g}^*$ level is very high in energy relative to the other ligand-field orbitals.

# BONDING IN TETRAHEDRAL COMPLEXES

11

As in crystal-field theory, we arrange the four ligands at the corners of a cube with the $x$-, $y$- and $z$-axes going through the cube faces. The σ-bonding ligand orbitals overlap in this case with the $d_{xy}$, $d_{yz}$ and $d_{xz}$ metal orbitals. In complexes of tetrahedral symmetry, these orbitals are labelled $t_2$ as you saw in Section 5.

⬤ Why is there no g subscript on this label?

⬤ Complexes in the point group $\mathbf{T_d}$ do not have a centre of symmetry. The subscript g refers to behaviour of the orbital when inverted through the centre of symmetry, and hence cannot be used for orbitals in tetrahedral complexes.

Because the σ-bonding orbitals do not point directly at the d orbitals (Figure 11.1) the σ-bonding in tetrahedral complexes is weaker than in octahedral complexes. Thus, the energy gap between the $t_2$ bonding and $t_2*$ antibonding orbitals is less than that between the $e_g$ bonding and $e_g*$ antibonding orbitals in octahedral complexes.

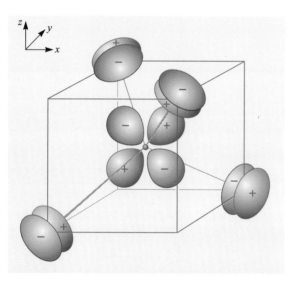

**Figure 11.1**  σ-bonding ligand 2p orbitals forming a tetrahedron around a metal $3d_{xz}$ orbital.

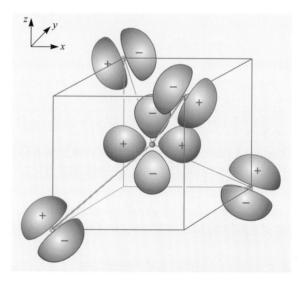

**Figure 11.2**  π-bonding ligand 2p orbitals overlapping with a metal $d_{x^2-y^2}$ orbital.

The π-bonding ligand orbitals overlap with the $d_{z^2}$ and $d_{x^2-y^2}$ metal orbitals. The complex orbitals formed will be labelled e. Figure 11.2 shows how π-bonding ligand orbitals overlap with a metal d orbital. Note that the overlap is not very different from that of the σ-bonding ligand orbitals with the metal d orbitals. Thus, the energy of the $t_2$ and e orbitals will not differ as much as the energy of the $t_{2g}$ and $e_g$ orbitals did for octahedral complexes.

Strong-field tetrahedral complexes are extremely rare as you saw in Section 5. We shall therefore only consider the case where both σ-bonding and π-bonding ligand orbitals are filled.

Each ligand has one σ-bonding orbital and two π-bonding orbitals available to overlap with the metal d orbitals. These orbitals are filled. Thus, there are twelve ligand orbitals providing a total of 24 electrons. Three of the σ-bonding orbitals combine with $d_{zy}$, $d_{yz}$ and $d_{xz}$ metal orbitals to form $t_2$ and $t_2^*$ orbitals. Two combinations of π-bonding ligand orbitals will overlap with $d_{z^2}$ and $d_{x^2-y^2}$ to form the e and e* orbitals. This leaves seven non-bonding combinations of ligand orbitals — one of σ-bonding orbitals and six of π-bonding orbitals. The orbital energy-level diagram for a weak-field tetrahedral complex is shown in Figure 11.3.

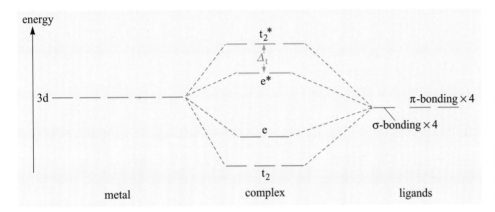

**Figure 11.3** Partial orbital energy-level diagram for a tetrahedral complex formed from filled σ-bonding and π-bonding ligand orbitals. Only orbitals of the complex with contributions from metal d orbitals are shown.

As usual, we do not show the non-bonding ligand orbitals. In the energy-level diagram, we have:

- the lowest-energy orbital is the $t_2$ bonding orbital, followed by the e bonding orbital. The electrons from the ligands can be assumed to fill these orbitals.

- Then there are the seven non-bonding combinations of ligand orbitals. These will also be filled by electrons from the ligand orbitals. For clarity, the levels corresponding to these orbitals are not shown in Figure 11.3.

- Above the ligand and metal orbitals are the e* and $t_2^*$ antibonding orbitals, separated by an energy $\Delta_t$. The metal d electrons can be thought of as going into these orbitals. Thus, the e* and $t_2^*$ orbitals play the role of e and $t_2$ in crystal-field theory.

The weaker σ-bonding between metal d orbitals and ligand orbitals is one factor in reducing the ligand field for tetrahedral complexes. There is a further factor which also has consequences for the intensity of d ↔ d transitions in tetrahedral environments. Figure 11.4 shows a 4p metal orbital and ligand σ-bonding orbitals.

○ Will the metal 4p orbital overlap with the σ-bonding ligands to form a bonding orbital?

● Yes. The combination shown in Figure 11.4 has all four σ-bonding ligand orbitals overlapping with lobes of the 4p orbital of the same sign.

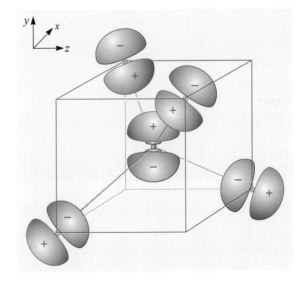

**Figure 11.4** 4p orbital on a metal and σ-bonding orbitals on tetrahedrally aranged ligands.

This overlap with the higher-energy 4p metal orbital will lower the energy of the $t_2$*
orbitals, thus reducing the ligand-field splitting even more. The orbital energy-level
diagram for a tetrahedral complex taking account of the involvement of the metal 4p
orbitals is shown in Figure 11.5.

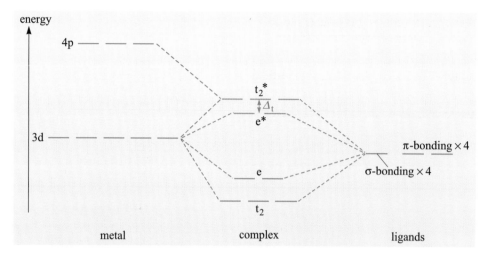

**Figure 11.5**   Partial orbital energy-level diagram for a tetrahedral complex formed from
filled σ-bonding and π-bonding ligand orbitals, showing the effect of involvement of the
metal 4p orbitals.

For tetrahedral complexes, the $p_x$, $p_y$ and $p_z$ metal orbitals have the same symmetry
label, $t_2$, as the $d_{xy}$, $d_{yz}$ and $d_{xz}$ orbitals. This coincidence of symmetry of p and d
orbitals has consequences for the spectra of tetrahedral complexes.

Because the $t_2$* complex orbitals now have some metal p character, the ligand-field
transition e* ↔ $t_2$* is now partly metal d ↔ p.

🔘  How will this affect the intensity of the ligand-field transition spectral band?

🔘  The transition will now be partly allowed because d ↔ p transitions are Laporte
allowed, so the spectral band will be more intense than in octahedral or square-
planar complexes.

Thus, molecular orbital theory can explain the weaker ligand field and more intense
visible spectra of tetrahedral complexes compared with octahedral complexes.

### QUESTION 11.1

What is the electronic configuration of the tetrahedral complex $[MnBr_4]^{2-}$?
Even though this complex is tetrahedral, its d ↔ d absorption spectrum is
very weak. Explain why.

### QUESTION 11.2

In octahedral symmetry should a 4p orbital on the transition-metal atom be
labelled with a subscript g or u? Will the 4p orbital contribute to the ligand-field
orbitals in octahedral complexes?

# 11.1 Summary of Section 11

1   Bonding is weaker in tetrahedral complexes than in octahedral complexes or those of $\mathbf{D}_{4h}$ symmetry because the ligand orbitals do not point directly at the metal orbitals.

2   $\sigma$-bonding ligand orbitals combine with the metal $d_{xy}$, $d_{yz}$ and $d_{xz}$ orbitals to form $t_2$ complex orbitals and $\pi$-bonding ligand orbitals combine with the $d_{z^2}$ and $d_{x^2-y^2}$ metal orbitals to form e orbitals.

3   Tetrahedral complexes are almost always weak field, and the ligand-field orbitals are e* and $t_2$*.

4   The higher-energy metal 4p orbital contributes to the $t_2$* orbital, thereby lowering it in energy.

5   The addition of p character to the $t_2$* orbital makes the $d \leftrightarrow d$ transitions in tetrahedral complexes more Laporte allowed than they are in octahedral complexes.

# COMPLEXES OF OTHER SYMMETRY

## 12

Most transition-metal complexes are less symmetrical than the ones we have been considering, and the d levels will split in more-complicated ways. However, if the deviations from high symmetry are not too severe, we can still use the diagrams we have obtained as a first approximation to those for the lower-symmetry complex. For example, a slightly distorted octahedral complex, or one with two *trans*-substituted ligands of a similar position in the spectrochemical series to the other four, will only show a small splitting *of* the $t_{2g}$ and $e_g$ levels relative to the splitting *between* the $t_{2g}$ and $e_g$ levels. To a first approximation, therefore, we could regard such complexes as octahedral. If a complex contains two or more types of ligand with very different ligand fields, then we need to consider the energy-level diagram appropriate to the actual symmetry of the complex.

An important type of substitution to consider is one in which a centre of symmetry is destroyed. Without a centre of symmetry, d $\leftrightarrow$ d transitions become partly allowed, and so the spectrum of such a complex is more intense than that of the unsubstituted complex.

### QUESTION 12.1

The intensity of the bands in the ligand-field spectrum of *cis*-$[CoF_2en_2]^+$ is greater than that of *trans*-$[CoF_2en_2]^+$. Explain why this is so.

# COMPLEXES WITH TWO METAL ATOMS

**13**

So far, we have considered metal complexes with a single metal atom surrounded by ligands, but there are many complexes with two or more metal centres, including some enzymes. These metal atoms may be bonded to each other, linked by ligand bridges or separated by several atoms. In the last case, the metal atoms can be considered to act separately. We shall briefly consider complexes with just two metal atoms (binuclear complexes) and with the metal atoms bonded together. Under metal–metal charge-transfer transitions in Section 14, we shall look at complexes with nearly independent atoms.

Think about what would happen if we had a simple diatomic molecule composed of two transition-metal atoms. We place the $z$-axis along the metal–metal bond, as we would for any diatomic molecule, and then consider the overlap of d orbitals on each atom. The $d_{z^2}$ orbitals will overlap to form $\sigma$ orbitals in the same way as $p_z$ orbitals do in molecules such as $N_2$. The two-metal-atom molecule has a centre of symmetry, and so the bonding orbital is labelled $\sigma_g$ and the antibonding orbital $\sigma_u$.

Figure 13.1 shows the $d_{xz}$ orbitals on each metal atom.

- What label would you give the bonding orbital formed from the $d_{xz}$ orbitals in Figure 13.1?

- Rotation of one half of a revolution about the molecular axis will produce an orbital of opposite sign to the original, making it a $\pi$ orbital. Inversion through the centre of symmetry also changes its sign. Hence the orbital is a $\pi_u$ orbital.

The antibonding combination will form a $\pi_g$ orbital. The $d_{yz}$ orbitals will also form $\pi$ orbitals, and these will be degenerate with those formed by the $d_{xz}$ orbitals.

The $d_{xy}$ and the $d_{x^2-y^2}$ metal orbitals also overlap, but these form a new type of orbital that we could not form by combining p orbitals. These orbitals are concentrated in planes perpendicular to the metal–metal bond and overlap face-on (Figure 13.2).

Rotation of these orbitals through half a revolution about the molecular axis produces an identical orbital, but rotation through one-quarter of a revolution changes the sign. These orbitals are

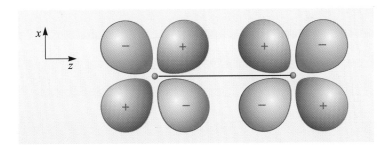

**Figure 13.1**  The bonding combination of $d_{xz}$ orbitals on two $\sigma$-bonded transition-metal atoms.

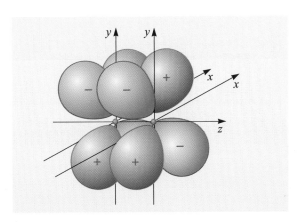

**Figure 13.2**  Two $d_{xy}$ orbitals overlapping face-on on two $\delta$-bonded transition-metal atoms.

**71**

labelled $\delta$ (delta). The bonding combinations are $\delta_g$ and the antibonding combinations are $\delta_u$. Like $\pi$ orbitals, $\delta$ orbitals come in degenerate pairs — that is, pairs of the same energy. The complete energy-level diagram for a diatomic transition-metal molecule using only the d orbitals is shown in Figure 13.3. Note that the strength of bonding is $\delta < \pi < \sigma$.

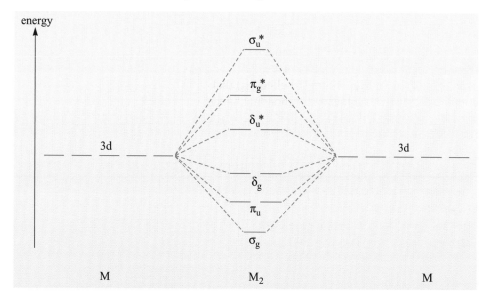

**Figure 13.3**  Energy-level diagram for a diatomic transition-metal molecule, $M_2$, showing levels for orbitals made by combining d orbitals only.

Now we can consider the effect of introducing ligands around the metal atoms. We shall take only one simple type of complex in which each metal atom has around it an approximately octahedral arrangement of atoms, as in the chromium(II) carboxylato complexes shown in Structure **13.1**. The other metal atom forms one vertex of the octahedron, and there is a ligand on each atom *trans* to this. The other four atoms coordinating to each metal form a square plane, and the two planes are in the same alignment. A number of chromium(II) complexes with this structure are known; one such complex was discovered as long ago as 1844. This complex puzzled the discoverers because it was red and diamagnetic, whereas most chromium(II) complexes are blue or violet and paramagnetic. The ligands in these complexes often coordinate through O or N and can be considered weak field.

**13.1**

⬤ What is the electronic configuration of $Cr^{2+}$?

◗ $3d^4$.

⬤ How many unpaired electrons would you expect in weak-field and strong-field octahedral complexes of chromium(II)?

◗ In a weak-field complex there would be four unpaired electrons ($t_{2g}^3 e_g^1$), and in a strong-field complex there would be two unpaired electrons ($t_{2g}^4 e_g^0$).

So if the two chromium atoms acted independently, we would expect the complexes to be paramagnetic. To explain the diamagnetism, we have to start from the bonding orbitals in Figure 13.3 ($\sigma_g$, $\pi_u$ and $\delta_g$). These combine with the ligand orbitals to form a set of orbitals in which metal and ligand orbitals are combined in a bonding manner, and one in which they are combined in an antibonding manner.

The $\sigma_g$ orbital in Figure 13.3 combines with $\sigma$-bonding ligand orbitals to form a bonding orbital, $a_{1g}$, and an antibonding combination, $a_{1g}^*$. The labels are those appropriate to the symmetry point group $D_{4h}$, to which complexes such as those we are discussing belong. Note that both $a_{1g}$ and $a_{1g}*$ orbitals are bonding between the two metal atoms; it is the overlap between metal and ligand orbitals which is antibonding in $a_{1g}*$. The $\pi_u$ orbitals overlap with axial $\pi$-bonding ligand orbitals to form $e_u$ and $e_u*$ orbitals.

In complexes of $D_{4h}$ symmetry the $\delta$ orbitals are no longer degenerate. You saw in Section 10 that $d_{x^2-y^2}$ orbitals overlap with $\sigma$-bonding ligand orbitals, whereas $d_{xy}$ orbitals overlap with $\pi$-bonding ligand orbitals. $\sigma$-bonding is generally much stronger than $\pi$-bonding, so that the $\delta_g$ orbital that is the bonding combination of $d_{x^2-y^2}$ orbitals combines with $\sigma$-bonding ligand orbitals to give an orbital (labelled $b_{1g}$) that is of lower energy than the bonding combination of $d_{xy}$ orbitals with $\pi$-bonding ligand orbitals (labelled $b_{2g}$). As a consequence, the $b_{1g}*$ level is at much higher energy than the $b_{2g}*$ level.

The electrons from the ligands are sufficient to fill the lowest bonding levels — $a_{1g}$, $e_u$, $b_{1g}$ and $b_{2g}$. Hence the electrons from the two metal ions can be allocated to the levels $a_{1g}*$, $e_u*$, $b_{2g}*$ and $b_{1g}*$. Typically, these orbitals will lie in the order just given, with the $b_{1g}*$ level very much higher in energy (in some complexes it is even higher than the orbitals formed from the antibonding $\pi_g*$ and $\delta_u*$ combinations of metal d orbitals). With four electrons from each chromium atom, there are just enough to fill the $a_{1g}*$, $e_u*$ and $b_{2g}*$ orbitals with all the electrons paired, as in Figure 13.4.

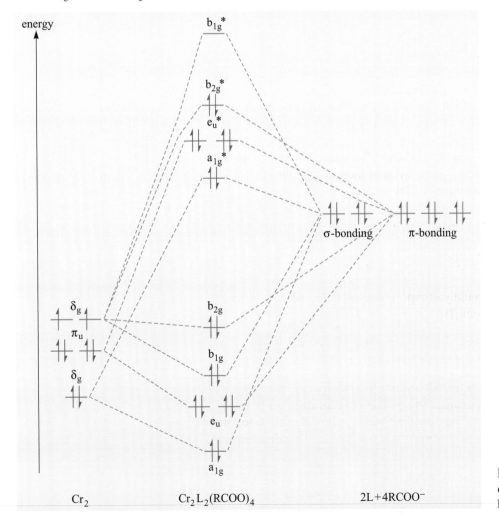

**Figure 13.4** Partial energy-level diagram for Structure **13.1**, which has $D_{4h}$ symmetry.

As for complexes of just one metal atom, the energy-level diagram only shows those levels involving metal d orbitals.

○ What will be the magnetism of the dichromium(II) complexes?

○ The complexes have all their electrons paired, and so will be diamagnetic in agreement with experimental observations.

QUESTION 13.1

From Figure 13.3, what would be the bond order of the molecule $Cu_2^{4+}$ formed by combining two $Cu^{2+}$ ions?

## 13.1 Summary of Section 13

1   d orbitals on two metal atoms can combine to form bonding and antibonding $\sigma$, $\pi$, and $\delta$ orbitals.

2   Orbitals for complexes with two metal atoms can be formed by combining the $\sigma$, $\pi$, and $\delta$ orbitals with ligand orbitals.

3   Well-known examples of complexes with two metal atoms are the set of chromium(II) carboxylato complexes. These are diamagnetic, unlike the paramagnetic mononuclear chromium(II) complexes.

# CHARGE-TRANSFER BANDS IN THE ELECTRONIC SPECTRA OF TRANSITION-METAL COMPLEXES

14

As we mentioned earlier, some of the deepest and most striking colours in transition-metal chemistry are not due to $d \leftrightarrow d$ electronic transitions, but to so-called *charge-transfer transitions*. Examples include the yellow of the chromate(VI) ion and orange of the dichromate(VI) ion (Figure 14.1), the deep purple of the manganate(VII) ion, the intense red of $[Fe(SCN)_4]^-$ and the orange of $TiBr_4$. The intensity of colour of some of these complexes such as $[Fe(SCN)_4]^-$ and $[Ni(dmg)_2]^{2- \dagger}$ accounts for their use in analysis. The colours of the pigments yellow ochre (hydrated iron(III) oxide) and Prussian blue ($KFe^{II}[Fe^{III}(CN)_6]$) are also due to charge-transfer transitions. Such transitions even occur in biological systems such as the blue copper proteins (Box 14.1, p. 79).

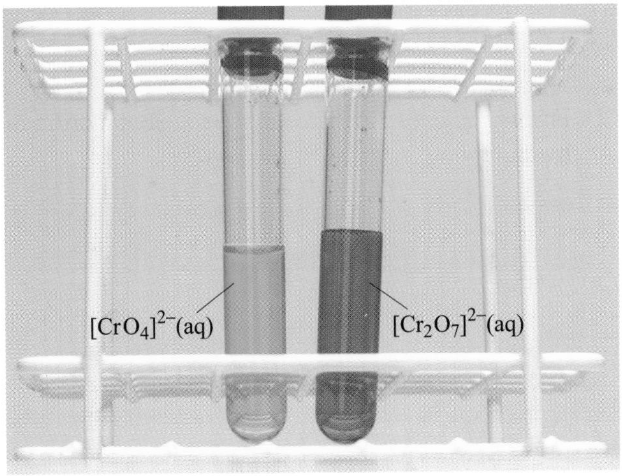

But what are the characteristics of such transitions, and from which electronic transitions do they arise?

Let us consider the following observations.

**Figure 14.1**  Solutions containing chromate(VI) and dichromate(VI) ions.

- Charge-transfer transitions are generally very intense; their molar absorption coefficients are of the order of $10^3$ to $10^4\, l\, mol^{-1}\, cm^{-1}$, in contrast to those of $d \leftrightarrow d$ transitions whose molar absorption coefficients are orders of magnitude less.
- The centre of the peak of a charge-transfer transition will lie at a shorter wavelength (higher wavenumber) than that of a $d \leftrightarrow d$ transition of the same complex.

In Section 3, you saw that $d \leftrightarrow d$ transitions were weak because they disobeyed the selection rules for electronic transitions.

🔵 What is the selection rule that $d \leftrightarrow d$ transitions break?

🔵 The Laporte selection rule, which states that the orbital quantum number, $l$, can only change by $\pm1$. Some very weak transitions also break the spin selection rule.

The spin selection rule still applies to charge-transfer transitions. The Laporte selection rule only applies if the electronic transition is between two metal-based orbitals. If one of the orbitals is based on the ligands, then it is inappropriate to assign it a metal orbital quantum number. For complexes with a single metal centre, it is transitions from a mainly metal orbital to a mainly ligand orbital or vice versa that give rise to charge-transfer spectra. If there is more than one metal atom in the complex, metal-to-metal transitions can also occur.

The name, **charge-transfer spectrum**, arises because an electron undergoing a transition giving rise to this type of spectrum goes from an orbital based on one atom to an orbital based on another. Three types of charge transfer can be distinguished: (i) ligand to metal, where the electron goes from a predominantly ligand orbital to a predominantly metal orbital; (ii) metal to ligand (metal orbital to ligand orbital); and (iii) metal to metal, where the electron goes from a metal atom in one oxidation state to a metal atom in another oxidation state (for example, iron(II) to iron(III)).

To see whether such a transition is allowed or forbidden, we need to use a new selection rule based on symmetry; that is, only transitions between levels with particular

$\dagger$ dmg is dimethylglyoxime.

symmetry labels are allowed. If we consider transitions of an electron to or from a $t_{2g}$ level in an octahedral complex, for example, it turns out that the following are allowed: $t_{2g} \leftrightarrow t_{1u}$, $t_{2g} \leftrightarrow t_{2u}$, $t_{2g} \leftrightarrow a_{2u}$. But what type of orbitals are labelled $t_{1u}$, $t_{2u}$ or $a_{2u}$? Well, p orbitals are labelled $t_{1u}$, and so combinations of ligand orbitals with symmetry similar to p orbitals will also be labelled $t_{1u}$. An example would be a ligand σ-bonding orbital on the z-axis in the positive direction with one sign, and a ligand σ-bonding orbital on the z-axis in the negative direction with the opposite sign (Figure 14.2).

This suggests that transitions from metal d orbitals to orbitals of the same symmetry as p orbitals are allowed. Recall that d ↔ p transitions are allowed by the Laporte selection rule.

⬤ What other transition from a d orbital is allowed by the Laporte rule?

⬤ d ↔ f. In both these situations, the orbital quantum number, $l$, changes by ±1.

In octahedral symmetry, f orbitals split into three levels labelled $a_{2u}$, $t_{1u}$ and $t_{2u}$. $a_{2u}$ and $t_{2u}$ combinations of ligand orbitals therefore have the same symmetry as metal f orbitals. Thus, although the Laporte selection rule does not apply, we can see some similarities to this selection rule in the symmetry-based selection rules. Transitions are allowed between metal d orbitals and combinations of ligand orbitals with the same symmetry as p and f orbitals. The allowed transitions for complexes of tetrahedral, $\mathbf{T_d}$, and $\mathbf{D_{4h}}$ symmetry also follow this pattern.

We shall now look at some examples of charge-transfer spectra.

# 14.1 Ligand-to-metal charge-transfer bands

In ligand-to-metal transitions, an electron jumps from an orbital of the complex mainly composed of ligand orbitals at low energy to one of the ligand-field orbitals in which there is a large contribution from the metal d orbitals. The ligand orbital will be one of those we have so far considered as non-bonding, and will have the same symmetry as a p or f metal orbital.

Interesting examples are the tetrahedral halide complexes. For such complexes, the ligand orbital is halogen $n$s or a combination of $n$s and $n$p orbitals. The promotion of electrons from halogen orbitals to the metal d orbitals gives rise to the charge-transfer spectra, and, for example, the orange colour of $TiBr_4$.

⬤ In titanium(IV) complexes how many electrons are there in the metal d orbitals?

⬤ There are no d electrons in titanium(IV) complexes.

The highest-occupied orbitals of the $TiX_4$ complexes are a non-bonding combination of π-bonding $n$p orbitals labelled $t_1$. This is the same symmetry as a metal f orbital. Figure 14.3 is an orbital energy-level diagram for $TiX_4$, which includes some of the ligand orbitals involved in charge-transfer transitions as well as the orbitals involving metal d orbitals.

⬤ Will a transition from this orbital to the ligand-field e* and $t_2$* orbitals be allowed according to the Laporte selection rule?

⬤ Yes. Transitions from ligand orbitals with the same symmetry as f orbitals, to metal d orbitals are allowed.

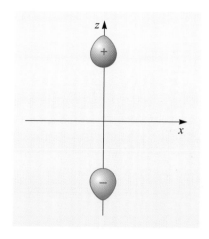

**Figure 14.2** A $t_{1u}$ combination of ligand σ-bonding orbitals.

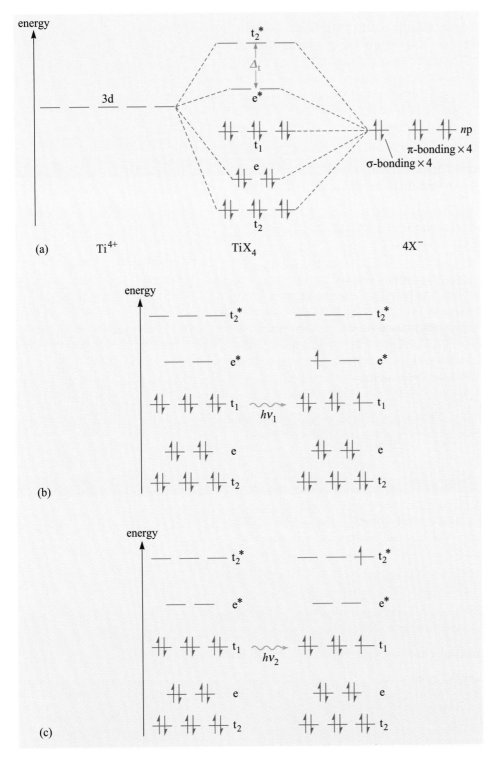

(a) Ti$^{4+}$     TiX$_4$     4X$^-$

(b)

(c)

**Figure 14.3** (a) Partial orbital energy-level diagram for TiX$_4$; (b) energy levels involved in the lowest-energy charge-transfer transition; (c) energy levels involved in the next highest-energy charge-transfer transition.

The lowest-energy transition ($h\nu_1$) shown in Figure 14.3b is from the TiX$_4$ non-bonding t$_1$ orbital to the e* orbital. The next highest-energy transition ($h\nu_2$) is either from the TiX$_4$ t$_1$ orbital to the t$_2$* orbital, or from the bonding t$_2$ orbital (which is more ligand than metal in character) to the e* orbital.

If the second transition is t$_1 \rightarrow$ t$_2$* (Figure 14.3c), what is the difference in energy ($h\nu_2 - h\nu_1$) between the first two transitions?

The ligand-field splitting energy, $\Delta_t$. In both transitions the electron starts in the $t_1$ level, but for the lower-energy one it ends up in $e^*$, and for the higher one it ends up in $t_2^*$. The energy difference between $e^*$ and $t_2^*$ is, of course, the ligand-field splitting energy, $\Delta_t$.

Thus, in some cases we can obtain ligand-field splitting energies from charge-transfer spectra.

The wavelengths of these spectra depend on the oxidisability of the ligand — in other words, the ease with which an electron can be removed from the ligand. The wavenumber of a particular ligand-to-metal charge-transfer band in halide complexes will increase with increase in the ionisation energy of the ligand $n$p electrons. This is because as the energy of the $n$p levels decreases, the ionisation energy increases, and so the energy gap between the ligand and metal orbitals increases. The ionisation energies increase in the order $I < Br < Cl$, so the wavenumber (and therefore energy) of the $t_1 \rightarrow e^*$ transition (Figure 14.3b) in $TiX_4$ also increases in this order. This transition is at $19\,600\ cm^{-1}$ for $TiI_4$, at $29\,500\ cm^{-1}$ for $TiBr_4$ and at $35\,400\ cm^{-1}$ for $TiCl_4$. The spectral bands corresponding to these transitions are very broad, and it is the spread into the blue–violet region that gives $TiBr_4$ its orange colour.

The nickel complexes $[Ni(NCO)_4]^{2-}$ and $[Ni(NCS)_4]^{2-}$ have spectral bands at $36\,300\ cm^{-1}$ and $26\,300\ cm^{-1}$, respectively, which have been assigned to ligand-to-metal transitions. Account for the difference in wavenumber.

$NCS^-$ will have a higher-energy filled $\pi$-bonding orbital than $NCO^-$ (because sulfur p atomic orbitals are of higher energy than oxygen p atomic orbitals).

Note that as for the halide complexes, the wavenumber increases with the ionisation energy ($O > S$) of the coordinating atom.

For most tetrahedral transition-metal complexes, the d orbitals are partly occupied. It therefore becomes more difficult to obtain the ligand-field splitting energy, since we have to allow for the interaction of the promoted electron ($e^*$ or $t_2^*$) with the electrons already occupying the mainly metal d orbitals. None the less, the shift in wavenumber of the charge-transfer spectral bands with the ionisation energy of the ligand is still clearly seen, as exemplified by the nickel(II) halide complexes in Figure 14.4.

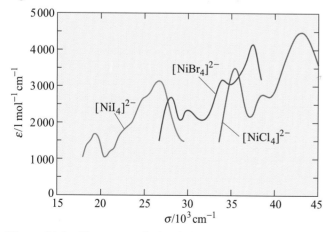

**Figure 14.4** Charge-transfer bands in the electronic absorption spectra of the complexes $[NiX_4]^{2-}$.

Which transition will give rise to the lowest-energy charge-transfer bands for the complexes whose spectra are shown in Figure 14.4?

$t_1 \rightarrow t_2{}^*$. Nickel(II) has an electronic configuration $d^8$. The $e^*$ level will therefore be full, and when excited the electron from the ligand $t_1$ level will have to go to the partly filled $t_2{}^*$ level.

## Box 14.1 A biological example of charge-transfer spectra—the blue copper proteins

Charge-transfer spectra can be the reason for colour in molecules of biological origin. An example of biological molecules whose colour arises from a ligand-to-metal charge-transfer transition is a class of proteins known as *blue copper proteins*. Although blue is the colour typically found for copper(II) compounds such as hydrated copper sulfate, the blue copper proteins are a much deeper blue, and the transition giving rise to the colour has a molar absorption coefficient too large to be that of a d $\leftrightarrow$ d transition. Azurin (Figure 14.5), plastocyanin and stellacyanin contain copper ions surrounded by a distorted tetrahedron of two sulfur atoms and two nitrogen atoms, which originate from cysteine and histidine residues, respectively, on the surrounding protein. Strong spectral bands arising from a transition from $\pi$-bonding orbitals on sulfur in a cysteine residue, to the empty $t_2{}^*$ on copper occur at wavenumbers in the range $12\,000\,\text{cm}^{-1}$ to $21\,000\,\text{cm}^{-1}$ (infrared to green).

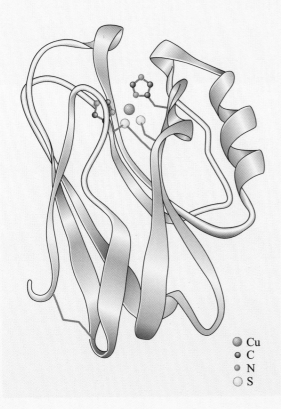

● Cu
● C
● N
○ S

**Figure 14.5** Schematic structure of azurin.

## QUESTION 14.1

Explain the changes in the wavenumber (Table 14.1) of the lowest-energy charge-transfer band ($t_1 \rightarrow e^*$) in the iron(III) complexes $[FeX_4]^-$, where X is a halogen.

In addition, the complex $[FeCl_4]^-$ has a band at $31\,500\ cm^{-1}$, which is assigned to the transition $t_1 \rightarrow t_2^*$. Use this information and that in the table above to estimate $\Delta_t$ for this complex.

**Table 14.1**

| Complex | Wavenumber of charge-transfer band/cm$^{-1}$ |
|---|---|
| $[FeCl_4]^-$ | 27 500 |
| $[FeBr_4]^-$ | 21 350 |
| $[FeI_4]^-$ | 14 300 |

## 14.2 Metal-to-ligand charge-transfer bands

If a complex has electrons in the metal d orbitals, they can be excited to unoccupied ligand orbitals lying at higher energy. Such a transition is a metal-to-ligand charge-transfer transition, as charge (the electron) has gone from a metal-based orbital to a ligand-based orbital.

⬤ Which ligands will give complexes with unoccupied orbitals lying not far above the metal d orbitals?

⬤ Those with unoccupied π orbitals at slightly higher energy than the metal d orbitals.

Charge-transfer bands of this type are thus characteristic of complexes containing strong-field ligands such as CO, CN$^-$ and PR$_3$. Ligands with delocalised π orbitals, such as phen and bipy, will also provide suitable empty π* orbitals. The intense red colour of bipy complexes of iron(II), for example, are due to a metal-to-ligand charge-transfer transition. For octahedral hexacyano complexes, the transitions are between the ligand-field orbitals, $t_{2g}$ and $e_g^*$, and the combinations of the unoccupied ligand π-bonding orbitals that we earlier considered as non-bonding. Thus, for $[Fe(CN)_6]^{4-}$, bands at $45\,780\ cm^{-1}$ and $50\,000\ cm^{-1}$ have been assigned to the transitions $t_{2g} \rightarrow t_{1u}$ and $t_{2g} \rightarrow t_{2u}$, respectively, where $t_{1u}$ and $t_{2u}$ are almost non-bonding combinations of $2\pi^*$ orbitals of the CN$^-$ ligands (Figure 14.6).

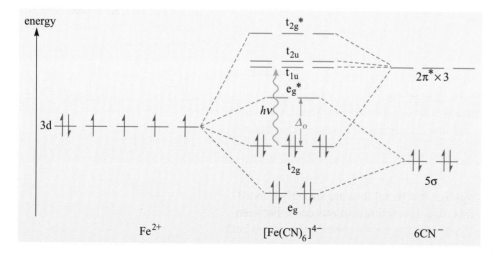

**Figure 14.6**  Partial orbital energy-level diagram for $[Fe(CN)_6]^{4-}$, showing the ligand orbitals involved in metal-to-ligand charge-transfer transitions.

QUESTION 14.2

Look back at the electronic spectra of titanium complexes in Figure 3.2. Identify the charge-transfer bands.

QUESTION 14.3

The complex $[Cr(CO)_6]$, which contains chromium in oxidation state 0 and can be considered a $d^6$ octahedral complex similar to $[Fe(CN)_6]^{4-}$ (Figure 14.6), has bands in its electronic spectrum at $29\,500\,cm^{-1}$ and $31\,550\,cm^{-1}$ of relatively low intensity, and a peak at $35\,700\,cm^{-1}$ with $\varepsilon = 13\,100\,l\,mol^{-1}\,cm^{-1}$. Suggest which types of transition give rise to these bands.

# 14.3 Metal-to-metal charge-transfer bands

When there are two or more metal centres close together, spectral transitions can occur between the orbitals based on one metal and those on the other. An interesting set of complexes is those with one metal in two different oxidation states. Such complexes are often intensely coloured. An example is the deep-blue compound known as Prussian blue, $KFe^{III}[Fe^{II}(CN)_6]$, in which the iron(III) ions are octahedrally surrounded by the nitrogen atoms of the $[Fe^{II}(CN)_6]^{4-}$ ions (Figure 14.7). The blue colour is due to transitions from a $t_{2g}$ orbital on iron(II) in the $[Fe^{II}(CN)_6]^{4-}$ ion to the $t_{2g}$ and $e_g^*$ orbitals on iron(III).

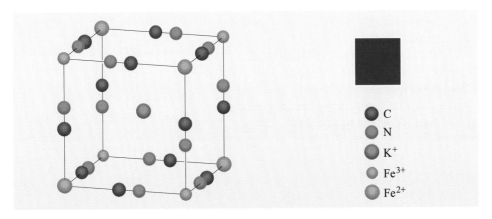

C
N
$K^+$
$Fe^{3+}$
$Fe^{2+}$

**Figure 14.7** The crystal structure of Prussian blue, $KFe^{III}[Fe^{II}(CN)_6]$. The square inset shows the colour of the compound.

○ Why are the transitions only from a $t_{2g}$ orbital of $[Fe^{II}(CN)_6]^{4-}$, and not from an $e_g^*$ orbital?

○ This complex will be strong field and iron(II) is $d^6$. In strong-field octahedral $d^6$ complexes, the electrons completely fill the $t_{2g}$ orbitals, and the $e_g^*$ orbitals are empty.

In $KFe^{III}[Fe^{II}(CN)_6]$, the $Fe^{III}$ and $Fe^{II}$ atoms essentially act independently of each other, but in some transition-metal complexes, the metal ions are strongly bonded together, as you saw in Section 12. In this case, spectral transitions occur between energy levels of the complex, for which the orbitals contain contributions from both metal ions. For the chromium complexes discussed in Section 13, transitions occur from the filled levels in Figure 13.4 to higher energy levels (not shown) for which the overlap of d orbitals on the two metal atoms is antibonding. The observed spectrum, giving rise to the red colour of the complex, has been interpreted in terms of transitions from $b_{2g}*$ to a $b_{1u}$ orbital from the $\delta*$ combination of metal ions, and to a transition from a mainly ligand orbital to the $e_g$ orbital. In effect, charge (the electron) is being transferred from the metal–metal bond to a metal–ligand bond.

## 14.4   Summary of Section 14

1   Charge-transfer bands arise from transitions from complex molecular orbitals whose main contributions come from orbitals on one atom to complex molecular orbitals, whose main contributions come from another atom. Thus, during the transition, charge is transferred from one atom to another.

2   Such transitions are not forbidden by the Laporte selection rule and can be very intense ($\varepsilon$ of the order of $10^4\,l\,mol^{-1}\,cm^{-1}$).

3   Three types of charge-transfer transition can be distinguished: ligand $\rightarrow$ metal, metal $\rightarrow$ ligand and metal $\rightarrow$ metal.

4   In ligand $\rightarrow$ metal transitions, an electron goes from a lower-energy ligand orbital with the same symmetry as a metal p or f orbital to one of the ligand-field orbitals on the metal.

5   The wavenumber of a particular ligand-to-metal charge-transfer band in halide and similar complexes will increase with an increase in the ionisation energy of the ligand $n$p electrons.

6   In metal $\rightarrow$ ligand transitions, an electron goes from one of the ligand-field orbitals on the metal to a higher-energy ligand orbital with the same symmetry as a metal p or f orbital.

7   In complexes with more than one metal atom, metal $\rightarrow$ metal transitions can occur in which the electron goes from an orbital based on a metal atom in one oxidation state to an orbital based on a metal atom in a different oxidation state.

8   Metal complexes of ligands with empty $\pi$-bonding ligands, such as $CN^-$, and ligands with delocalised $\pi$ orbitals, often have intense metal $\rightarrow$ ligand charge-transfer bands.

# REVISION EXERCISE: COMPLEXES OF COBALT(II) AND COBALT(III)

# 15

The questions in this section cover the main points of the text. You can read these and their answers as a summary, try the questions now to check that you have understood the main points we have covered, or save them for revision later.

**Q1** $Co^{2+}$ is a $d^7$ ion and $Co^{3+}$ is a $d^6$ ion. Sketch crystal-field energy-level diagrams for octahedral complexes of these ions, showing only the metal 3d levels, and filling in the electrons for the strong-field and weak-field cases.

**Q2** The ionic radius of $Co^{2+}$ in crystals with high-spin (weak-field) octahedral environments (for example, $CoF_2$) is smaller than expected for an ion in which the electrons are equally distributed between all five 3d orbitals (a spherical crystal field). Explain this smaller radius using crystal-field theory and molecular orbital theory.

**Q3** The lattice energy of $CoCl_2$ is less than that predicted for an ion in a spherical crystal field with the same number of d electrons. Explain this difference using crystal-field theory.

**Q4** In solution, the ion $[Co(PhCN)_6]^{2+}$, containing the ligand PhCN, is a distorted octahedral complex belonging to the symmetry point group $\mathbf{D}_{4h}$. PhCN is a strong-field ligand. Explain why you would expect the complex to be distorted from a true octahedron. Is it possible to predict whether the complex will have two bonds longer than the other four or two bonds shorter than the other four?

**Q5** Square-planar complexes of cobalt(II) have magnetic moments in the range $2.2$–$2.7\,\mu_B$. Explain why these values are much lower than those for tetrahedral complexes of this ion ($4.4$–$4.8\,\mu_B$).

**Q6** Sketch a $t_{2g}$ orbital and a $t_{2g}{}^*$ orbital formed from a cobalt $3d_{xy}$ orbital and empty $\pi$-bonding orbitals on ligands such as CO.

**Q7** Sketch a partial orbital energy-level diagram for octahedral $[Co(CN)_6]^{3-}$, showing molecular orbitals formed from $CN^-$ orbitals and only the 3d orbitals on cobalt. Show on your diagram how electrons occupy the available levels.

**Q8** If a metal cation replaces $Li^+$ in a lithium halide crystal as an isolated impurity, then the impurity cation can be regarded as surrounded by an octahedron of halide ions to which it is bonded. For $Co^{2+}$ impurities in LiCl and LiBr, spectral bands arising from the transition from a combination of $\pi$-bonding ligand orbitals on the halogen ligands to the cobalt $t_{2g}{}^*$ level are observed at $39\,400$ and $30\,600\,cm^{-1}$, respectively. Bands for the same system from the transition from the same $\pi$-bonding ligand orbital combination on the halogen atoms to the cobalt $e_g{}^*$ level are observed at $45\,000$ and $35\,300\,cm^{-1}$, for LiCl and LiBr, respectively. Use these observations to estimate $\Delta_o$ for $Co^{2+}$ in an octahedron of $Cl^-$ ions and in an octahedron of $Br^-$ ions. Are the variations in $\Delta_o$ and in the position of the spectral bands from $Cl^-$ to $Br^-$ as you would expect?

**Q9** Explain how the order of the ligand-field energy levels for *trans*-$[Co(NH_3)_2Cl_4]^-$ will differ from that in Figure 10.4. Given that the complex is low spin, what is its electronic configuration?

# APPENDIX   FLOW CHART FOR DETERMINING THE SYMMETRY POINT GROUP OF AN OBJECT

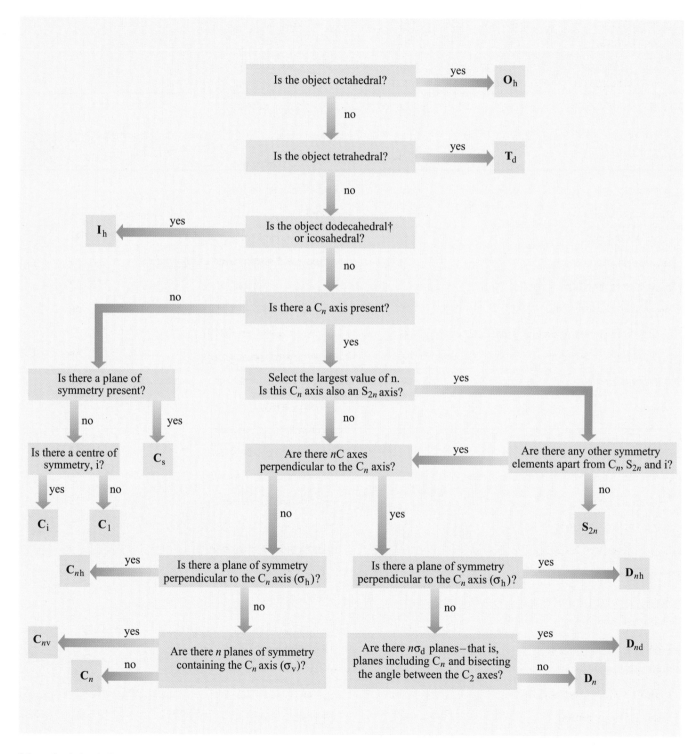

† A regular dodecahedron has twelve faces, each with five sides. The icosahedral group is relatively rare. The icosahedron has twenty triangular faces.

# LEARNING OUTCOMES

Now that you have completed *Metal–Ligand Bonding*, you should be able to do the following things:

1 Recognise valid definitions of, and use in a correct context, the terms, concepts and principles in the following Table. (All Questions)

List of scientific terms, concepts and principles introduced in *Metal–Ligand Bonding*

| Term | Page number | Term | Page number |
|---|---|---|---|
| barycentre | 6 | Laporte selection rule | 18 |
| Bohr magneton, $\mu_B$ | 34 | ligand | 1 |
| boundary surface | 2 | ligand-field splitting energy | 47 |
| charge-transfer spectra | 75 | ligand-field stabilisation energy, LFSE | 47 |
| coordination number | 1 | low spin | 7 |
| crystal-field splitting energy, $\Delta$ | 6 | magnetic moment, $\mu$ | 34 |
| crystal-field stabilisation energy, CFSE | 11 | magnetic susceptibility, $\chi$ | 33 |
| crystal-field theory | 5 | $\mathbf{O_h}$ symmetry point group | 59 |
| Curie constant | 34 | orbital magnetic moment | 38 |
| Curie's law | 34 | pairing energy, $P$ | 7 |
| d orbitals | 1 | paramagnetic substance | 33 |
| d $\leftrightarrow$ d or d–d spectra | 15 | $\pi$-bonding ligand orbital | 47 |
| $\delta$ orbitals | 72 | $\sigma$-bonding ligand orbital | 44 |
| degenerate state | 6 | spectrochemical series | 15 |
| diamagnetic substance | 33 | spin magnetic moment, $\mu_S$ | 36 |
| $\mathbf{D_{4h}}$ symmetry point group | 21 | spin-forbidden transition | 19 |
| e level | 28 | spin selection rule | 19 |
| $e_g$ set | 22 | state | 18 |
| electronic transition | 14 | strong field | 7 |
| g factor | 36 | $\mathbf{T_d}$ symmetry point group | 59 |
| high spin | 7 | $t_2$ level | 28 |
| improper axis of rotation, $S_n$ | 59 | $t_{2g}$ set | 14 |
| Jahn–Teller distortion | 23 | vibronic coupling | 20 |
| Jahn–Teller theorem | 23 | weak field | 7 |

2   Apply crystal-field theory and/or simplified molecular-orbital theory to a complex or molecule of given symmetry to explain the splitting of the d orbital energy levels. (Questions 2.1, 2.2, 5.2, 10.2, 10.3 and 11.2; Revision Exercise Q1, Q2, Q3, Q4, Q5, Q6, Q7 and Q9)

3   Calculate the CFSE or LFSE of tetrahedral and octahedral complexes, and use these to explain trends in properties across the first-row transition series. Relate the value of $\Delta$ to the position of the ligand in the spectrochemical series, the position of the metal in the Periodic Table and its oxidation state. (Questions 2.1, 2.2, 2.3, 2.4, 2.5, 3.1 and 5.2; Revision Exercise Q2 and Q3)

4   Relate the absorption spectrum of complexes to the appropriate energy-level diagram. (Questions 14.1 and 14.3; Revision Exercise Q7)

5   Predict whether a given transition-metal complex is likely to be tetrahedral, octahedral or square planar, and predict which complexes are likely to suffer large Jahn–Teller distortions. (Question 4.1; Revision Exercise Q4)

6   Write down the electronic configuration of complexes using crystal-field theory or ligand-field theory. (Questions 5.1, 9.1, 9.3, 10.2 and 11.1; Revision Exercise Q5 and Q9)

7   Relate electronic configuration to magnetic properties of metal complexes. (Questions 6.1, 6.2 and 6.3)

8   Recognise the presence of symmetry elements including $S_n$ axes in given molecules or other objects, and use a flow chart to determine the symmetry point group of a complex . (Questions 10.1, 11.2 and 12.1)

9   Use the concepts of $\sigma$ and $\pi$ bonding to explain orbital energy-level diagrams for transition-metal complexes, and rationalise the position of ligands in the spectrochemical series. (Questions 9.2, 9.4, 10.3 and 11.2; Revision Exercise Q7, Q8 and Q9)

10  Account for the intensities of bands in the electronic spectra of transition-metal complexes in terms of selection rules. (Questions 3.2, 3.3, 11.1 12.1, 14.2 and 14.3)

11  Explain why charge-transfer spectra are more intense that d $\leftrightarrow$ d spectra, and recognise different types of charge-transfer band. (Questions 14.1, 14.2 and 14.3; Revision Exercise Q8)

12  Apply the concepts of molecular orbital theory to complexes with two metal atoms. (Question 13.1)

# ANSWERS TO QUESTIONS

## QUESTION 2.1 (Learning Outcomes 2 and 3)

The $d^1$ to $d^3$ ions ($Sc^{II}$ to $V^{II}$) would behave exactly as for a weak-field compound such as $MF_2$. However, for a $d^4$ ion ($Cr^{II}$), there would now be four electrons in $t_{2g}$, and so the deviation would increase. Similarly, $d^5$ and $d^6$ ions ($Mn^{II}$ and $Fe^{II}$) would be smaller than in the $MF_2$ case, because only $t_{2g}$ orbitals would be occupied. At $d^7$ ($Co^{II}$), there is one electron in $e_g$, and so the deviation is less. The ionic radii would gradually approach the line expected for spherical ions through $d^8$ and $d^9$ until $d^{10}$ ($Zn^{II}$, spherical ion), is reached. The plot would therefore be expected to be a single bowl, with the largest deviation from the spherical ion depiction occurring at $d^6$. A sketch is given in Figure Q.1.

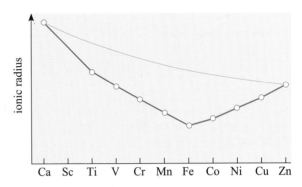

**Figure Q.1** Sketch of the predicted ionic-radii variation for divalent ions of calcium and the first-row transition metals in the dicyanides, $M(CN)_2$.

## QUESTION 2.2 (Learning Outcomes 2 and 3)

As in the case of the difluorides, the ionic radius of $M^{2+}$ in the oxides MO will deviate from the smooth curve through $Ca^{2+}$, $Mn^{2+}$ and $Zn^{2+}$. The deviation will be greatest for the $d^3$ and $d^8$ ions, $V^{2+}$ and $Ni^{2+}$. Your sketch should look like Figure 2.8.

In reality, the experimental curve would have further deviations from the ideal smooth curve because of other factors. Most of the known oxides MO have the NaCl structure. However, CuO contains $Cu^{2+}$ in a square-planar environment. In addition, TiO and VO have a degree of metal–metal bonding. The structure of CrO is not known.

## QUESTION 2.3 (Learning Outcome 3)

In a weak octahedral field, the CFSEs of the $M^{3+}$ ions are:

| $Sc^{3+}$ | $Ti^{3+}$ | $V^{3+}$ | $Cr^{3+}$ | $Mn^{3+}$ | $Fe^{3+}$ | $Co^{3+}$ | $Ni^{3+}$ | $Ga^{3+}$ |
|---|---|---|---|---|---|---|---|---|
| 0 | $\frac{2}{5}\Delta_o$ | $\frac{4}{5}\Delta_o$ | $\frac{6}{5}\Delta_o$ | $\frac{3}{5}\Delta_o$ | 0 | $\frac{2}{5}\Delta_o$ | $\frac{4}{5}\Delta_o$ | 0 |

The point for the $FeF_3$ lattice energy will lie on a smooth curve going through the points for $ScF_3$ and $GaF_3$. The points for the other fluorides will lie below this line. $CrF_3$ will lie most below, followed by $VF_3$ and $NiF_3$, $MnF_3$ and, lastly, $TiF_3$ and $CoF_3$. A rough sketch is shown in Figure Q.2. $CuF_3$ and $ZnF_3$ have not yet been synthesised.

You may be surprised to see $NiF_3$ included in the question. The dominant oxidation state of nickel is +2, but nickel(III) trifluoride has been synthesised as a black solid by the thermal decomposition of $NiF_4$. Copper(III) is also known in complexes but not as the trifluoride.

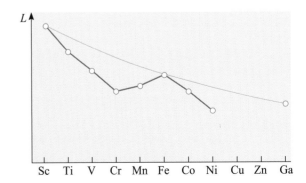

**Figure Q.2** Sketch of the predicted variation of the lattice energy of the metal trifluorides across the first transition series and gallium.

QUESTION 2.4 (*Learning Outcome 3*)

The CFSEs for the dicyanides will be:

$Ti^{2+}$ ($d^2$)  $\quad$ $V^{2+}$ ($d^3$) $\quad$ $Cr^{2+}$ ($d^4$) $\quad$ $Mn^{2+}$ ($d^5$) $\quad$ $Fe^{2+}$ ($d^6$) $\quad$ $Co^{2+}$ ($d^7$) $\quad$ $Ni^{2+}$ ($d^8$) $\quad$ $Cu^{2+}$ ($d^9$) $\quad$ $Zn^{2+}$ ($d^{10}$)

$\frac{4}{5}\Delta_o$ $\qquad$ $\frac{6}{5}\Delta_o$ $\qquad$ $\frac{8}{5}\Delta_o - P$ $\quad$ $2\Delta_o - 2P$ $\quad$ $\frac{12}{5}\Delta_o - 2P$ $\quad$ $\frac{9}{5}\Delta_o - P$ $\quad$ $\frac{6}{5}\Delta_o$ $\qquad$ $\frac{3}{5}\Delta_o$ $\qquad$ $0$

With $P = \frac{1}{2}\Delta_o$, the deviations from the smooth curve through $Ca^{2+}$ and $Zn^{2+}$ will be:

$Ti^{2+}$ $\frac{4}{5}\Delta_o$, $V^{2+}$ $\frac{6}{5}\Delta_o$, $Cr^{2+}$ $\frac{11}{10}\Delta_o$, $Mn^{2+}$ $\Delta_o$, $Fe^{2+}$ $\frac{7}{5}\Delta_o$, $Co^{2+}$ $\frac{13}{10}\Delta_o$, $Ni^{2+}$ $\frac{6}{5}\Delta_o$, $Cu^{2+}$ $\frac{3}{5}\Delta_o$, $Zn^{2+}$ $0$

The predicted curve is sketched in Figure Q.3.

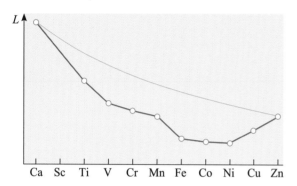

**Figure Q.3** Sketch of the predicted variation of lattice energies of $M(CN)_2$ salts from calcium to zinc.

QUESTION 3.1 (*Learning Outcomes 3 and 4*)

The spectrochemical series for the ligands in Table 3.2 is

$\quad$ $F^- < S_2PF_2^- < ox < CNO^- < CN^-$

$F^-$, ox and $CN^-$ are in the general series given in the margin of p. 15 in this order. $CNO^-$ is at the strong-field end, suggesting binding through C; $S_2PF_2^-$ is a weak-field ligand. You should assume that ligands binding through P occur at the strong-field end. $S_2PF_2^-$ binds through S, and such ligands are weak field; another example is $SCN^-$.

QUESTION 3.2 (*Learning Outcomes 4 and 10*)

The green colour of $[Ni(H_2O)_6]^{2+}$(aq) arises from d–d transitions of the metal ion. These are Laporte forbidden as $\Delta l = 0$. However, electronic excitation from $t_{2g}$ to $e_g$ (that is, $t_{2g}^6 e_g^2 \rightarrow t_{2g}^5 e_g^3$) *can* take place without breaking the spin selection rule ($\Delta s = 0$). d–d electronic transitions in $[Ni(H_2O)_6]^{2+}$(aq) are therefore Laporte forbidden, but spin allowed.

QUESTION 3.3 (*Learning Outcomes 4 and 10*)

Given that the glass contains $Fe^{3+}$ ($3d^5$) as an impurity, it is highly likely that the green colour arises from an iron d–d transition. If we assume that this ion is in an octahedral environment, a spin-forbidden $t_{2g}^3 e_g^2 \rightarrow t_{2g}^2 e_g^3$ transition would give rise to a low intensity colour. The fact that it is more clearly seen by viewing the bottle along its axis rather than side on is simply a consequence of the longer pathlength in the former case.

## QUESTION 4.1 (*Learning Outcome 5*)

You should recall that a Jahn–Teller distortion occurs when orbitals of equal energy (degenerate) are unequally populated. This is *only* the case for complex (b). In $[MnCl_6]^{3-}$, the metal is in the +3 oxidation state, giving a configuration $3d^4$. In a high-spin octahedral complex, the configuration will be $t_{2g}^3 e_g^1$, and a Jahn–Teller distortion is predicted.

## QUESTION 5.1 (*Learning Outcome 6*)

In the complex $[CoCl_4]^{2-}$, cobalt is in the +2 oxidation state, which means that there are seven d electrons. As the complex is tetrahedral, the configuration will be $e^4 t_2^3$, so that there are three unpaired electrons.

## QUESTION 5.2 (*Learning Outcome 3*)

(a)   The variation will be double-bowl shaped, with a smooth curve passing through the points $Ca^{2+}$, $Mn^{2+}$ and $Zn^{2+}$, and values below the curve for the other ions. The deviation from the curve depends on the CFSE in the tetrahedral complex. For a tetrahedral high-spin complex, the CFSE values are as in Table 5.1:

| $d^1$ | $d^2$ | $d^3$ | $d^4$ | $d^5$ | $d^6$ | $d^7$ | $d^8$ | $d^9$ | $d^{10}$ |
|---|---|---|---|---|---|---|---|---|---|
| $\frac{3}{5}\Delta_t$ | $\frac{6}{5}\Delta_t$ | $\frac{4}{5}\Delta_t$ | $\frac{2}{5}\Delta_t$ | 0 | $\frac{3}{5}\Delta_t$ | $\frac{6}{5}\Delta_t$ | $\frac{4}{5}\Delta_t$ | $\frac{2}{5}\Delta_t$ | 0 |

This gives a sketch of the form of Figure Q.4, with minima at titanium and cobalt. (Remember that virtually all tetrahedral complexes are high spin.)

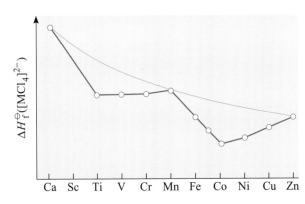

**Figure Q.4**   Predicted variation in $\Delta H_f^{\ominus}$ for tetrahedral $[MCl_4]^{2-}$ from calcium to zinc.

(b)   You should recall that the crystal-field view of transition-metal complexes compares the situation in the free ion to that in the complex. We cannot use the ion in the aqueous state in Equation 5.2, because in this state the ion is not 'free' but a hexaaquo complex, $[M(H_2O)_6]^{2+}$.

## QUESTION 6.1 (*Learning Outcome 7*)

(a)   The configuration of high-spin manganese(II) complexes is $t_{2g}^3 e_g^2$. Even in the free ion, there is no orbital contribution to the magnetic moment. Forming a complex will not introduce an orbital contribution, and so values close to $\mu_S$ are expected.

(b)   Those configurations with one, two, four or five electrons in $t_{2g}$. For strong-field complexes, these are $d^1$ $(t_{2g}^1)$, $d^2$ $(t_{2g}^2)$, $d^4$ $(t_{2g}^4)$, $d^5$ $(t_{2g}^5)$. In these configurations, we can have vacancies in $d_{xz}$ or $d_{yz}$ and a possible orbital contribution.

## QUESTION 6.2 (Learning Outcome 7)

$[NiF_6]^{2-}$ contains nickel in oxidation state +4, in which it has six 3d electrons. For an octahedral complex, the configuration would therefore be $t_{2g}^4 e_g^2$ (weak field) or $t_{2g}^6$ (strong field). $t_{2g}^4 e_g^2$ will have four unpaired electrons, and $t_{2g}^6$ none. The observed magnetic moment of $0\mu_B$ therefore implies a strong-field complex. You may be surprised that a fluoride complex can be a strong field, but $\Delta_o$ will be larger for nickel(IV) than for the common lower oxidation states. Recall that for a given ligand and a given metal, $\Delta$ increases with the oxidation state of the metal.

## QUESTION 6.3 (Learning Outcome 7)

The ions $[CoCl_4]^{2-}$ and $[Co(H_2O)_6]^{2+}$ both contain cobalt in oxidation state +2. These are therefore $d^7$ complexes. For tetrahedral complexes, we expect to find three unpaired electrons, and for octahedral complexes we expect to find three (weak field) or one (strong field) unpaired electrons.

The magnetic moments suggest three unpaired electrons, giving electronic configurations of $e^4 t_2^3$ (tetrahedral) and $t_{2g}^5 e_g^2$ (octahedral). Thus, the octahedral complex is weak field. The octahedral complex has a higher magnetic moment, because its orbital contribution is larger due to the unoccupied $t_{2g}$ level.

## QUESTION 9.1 (Learning Outcome 6)

For σ-bonded complexes, the six metal d electrons can be thought of as filling the $t_{2g}$ and $e_g^*$ levels. For strong-field $d^6$ complexes the electronic configuration is $t_{2g}^6$ and for weak-field complexes $t_{2g}^4 e_g^{*2}$.

## QUESTION 9.2 (Learning Outcome 9)

There are no 2d orbitals, and so nitrogen ligands $NR_3$ do not have empty d orbitals close in energy to the metal d orbitals. Phosphorus and arsenic both have available d orbitals, so that ligands coordinating to the metal through these atoms can form π bonds, which lead to the lowering of the $t_{2g}$ level. This increases the ligand-field splitting, giving a stronger field than for complexes with $NR_3$ ligands.

## QUESTION 9.3 (Learning Outcome 6)

(a) Weak field $t_{2g}^3 e_g^{*2}$, strong field $t_{2g}^5$; (b) $t_{2g}^5$; (c) $t_{2g}^{*3} e_g^{*2}$.

## QUESTION 9.4 (Learning Outcome 9)

$N^{3-}$ will have filled 2p orbitals. These will form fairly strong π bonds with metal d orbitals, thus raising the energy of the first level into which the metal d electrons are fed, the $t_{2g}^*$. The small $t_{2g}^*$–$e_g^*$ energy gap will make the ligand weak field. The 2p orbitals on $NH_3$ will bond less strongly with the metal $t_{2g}$ orbitals as they are involved in the N—H bonds; consequently the electron density is drawn away from the metal. The $t_{2g}^*$ level will therefore be less raised in energy than for $N^{3-}$, and the $t_{2g}^*$–$e_g^*$ energy gap will be larger. $NH_3$ is thus a stronger-field ligand than $N^{3-}$.

## QUESTION 10.1 (Learning Outcome 8)

(a) trans-$[FeCl_2Br_4]^{4-}$ is not truly octahedral; neither is it tetrahedral, dodecahedral or icosahedral. It has a $C_4$ axis (through ClFeCl) and $4C_2$ axes. The largest value of $n$ is 4. The complex has $4C_2$ axes perpendicular to the $C_4$ axis. There is a plane of symmetry perpendicular to the $C_4$ axis (the plane containing Fe and the four Br atoms). This complex belongs to the symmetry point group $\mathbf{D}_{4h}$.

(b) *cis*-[FeCl$_2$Br$_4$]$^{4-}$ is also not truly octahedral. Also it isn't tetrahedral, icosahedral or dodecahedral. It has a C$_2$ axis (bisecting the ∠ClFeCl angle), but no other axes of symmetry. The C$_2$ axis is not an S$_4$ axis. There are no C$_2$ axes perpendicular to the C$_2$ axis, and no plane of symmetry perpendicular to the C$_2$ axis. There are two vertical planes of symmetry — one containing the two Cl atoms and the two Br atoms *trans* to them, and one bisecting the ∠ClFeCl angle and containing the other two Br atoms. This complex belongs to the symmetry point group **C$_{2v}$**.

Such complexes as these are often regarded as roughly octahedral, but in accurate energy-level diagrams the levels of *trans*-[FeCl$_2$Br$_4$]$^{4-}$ will be labelled as in Section 10.2, but those of *cis*-[FeCl$_2$Br$_4$]$^{4-}$ will have labels similar to those used in Figure 9.17 for the water molecule (a$_1$, a$_2$, b$_1$ or b$_2$).

## QUESTION 10.2 (Learning Outcomes 2 and 9)

*trans*-[FeCl$_2$(H$_2$O)$_4$] contains iron in oxidation state +2. There are therefore six d electrons to feed into the ligand-field levels. Cl$^-$ is a weaker-field ligand than H$_2$O, so the diagram in Figure 10.4 is appropriate. The relevant levels to fill are e$_g$*, b$_{2g}$*, a$_{1g}$* and b$_{1g}$*. Thus, the electronic configuration is e$_g$*$^3$b$_{2g}$*$^1$a$_{1g}$*$^1$b$_{1g}$*$^1$.

## QUESTION 10.3 (Learning Outcome 9)

A major difference is that the platinum and palladium complexes contain weak-field ligands such as Br$^-$ with filled π-bonding ligand orbitals, whereas the nickel(II) complexes have empty π-bonding orbitals. For platinum and palladium complexes, b$_{2g}$ and e$_g$ will be full, and the metal d electrons will be allocated to the a$_{1g}$*, b$_{2g}$*, e$_g$* and b$_{2g}$* levels. In addition, the filled π-bonding ligand orbitals are at the same energy as the σ-bonding ligand orbitals so that the e$_g$* and b$_{2g}$* levels drop below b$_{1g}$*.

## QUESTION 11.1 (Learning Outcomes 6 and 10)

The complex contains manganese(II) and so has five d electrons. Its configuration is e*$^2$t$_2$*$^3$, as tetrahedral complexes are almost invariably weak field and Br$^-$ is a weak-field ligand.

An electron in a high-spin d$^5$ complex cannot undergo a transition from e* to t$_2$* without changing its spin. This is forbidden by the spin selection rule.

## QUESTION 11.2 (Learning Outcomes 8 and 9)

The 4p orbital changes sign when inverted through the centre of symmetry and so will be labelled with a subscript u. The ligand-field orbitals are labelled e$_g$ and t$_{2g}$ and only orbitals with the same symmetry (in particular only orbitals labelled with a subscript g) can combine with them. Hence the metal 4p orbital cannot contribute to the ligand-field orbitals in octahedral complexes.

## QUESTION 12.1 (Learning Outcomes 8 and 10)

*cis*-[CoF$_2$en$_2$]$^+$ does not have a centre of symmetry, so that the selection rule g ↔ g does not apply and d ↔ d transitions are partly allowed. *trans*-[CoF$_2$en$_2$]$^+$ does have a centre of symmetry, so d ↔ d transitions are forbidden and the spectral lines are weaker.

## QUESTION 13.1 (Learning Outcome 12)

With nine d electrons each from the two Cu$^{2+}$ ions, there are just enough to fill all the orbitals shown except the σ$_u$*. The bond order is thus

  (number of filled bonding orbitals)  − (number of filled antibonding orbitals) = 5 − 4 = 1

## QUESTION 14.1 (Learning Outcome 11)

The wavenumber of the $t_1 \to e^*$ transition in such complexes increases in the order of the ionisation energy of the ligand, I < Br < Cl.

To a first approximation, the difference in energy between the transitions $t_1 \to e^*$ and $t_1 \to t_2^*$ is the ligand-field splitting energy $\Delta_t$, which is equal to the $e^*$–$t_2^*$ energy gap. For $[FeCl_4]^-$, this is $31\,500 - 27\,500\,\text{cm}^{-1} = 4\,000\,\text{cm}^{-1}$.

## QUESTION 14.2 (Learning Outcomes 10 and 11)

The electronic spectra of $TiCl_3$ (Figure 3.2c) and $TiBr_3$ (Figure 3.2d) exhibit strong charge-transfer bands at higher wavenumber than the weak $d \leftrightarrow d$ bands. The very high intensity to the right of these spectra is part of the charge-transfer band.

## QUESTION 14.3 (Learning Outcomes 4, 10 and 11)

$[Cr(CO)_6]$ contains the strong-field ligand, CO, and so the highest-occupied level will be $t_{2g}$. The weak-intensity transitions are likely to be due to the $d \to d$ transition $t_{2g} \to e_g^*$. The peak with $\varepsilon = 13\,100\,1\,\text{mol}^{-1}\,\text{cm}^{-1}$ will be a charge-transfer transition, probably $t_{2g} \to t_{1u}$ or $t_{2g} \to t_{2u}$.

# ANSWERS TO REVISION EXERCISE QUESTIONS

### Q1 (Learning Outcome 2)

Figure Q.5 shows the crystal-field energy-level diagrams for octahedral $d^6$ and $d^7$ complexes.

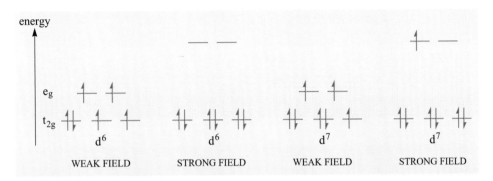

**Figure Q.5** Crystal-field energy-level diagrams for $d^6$ and $d^7$ ions in weak and strong fields for octahedral cobalt complexes.

### Q2 (Learning Outcomes 2 and 3)

The orbitals pointing towards the ligands (that is, $d_{z^2}$ and $d_{x^2-y^2}$) shield the ligands from the nucleus more than do the $d_{xy}$, $d_{yz}$ and $d_{xz}$ orbitals. Therefore, if there are more partially filled orbitals in the $e_g$ set ($d_{z^2}$ and $d_{x^2-y^2}$) than in the $t_{2g}$ set ($d_{xy}$, $d_{yz}$ and $d_{xz}$), the ligands can approach the metal more closely than they could for a spherical ion in which the electrons are evenly distributed between the sets. $Co^{2+}$ in a weak octahedral field has a configuration $t_{2g}^5 e_g^2$. It thus has one half-filled $t_{2g}$ orbital and two half-filled $e_g$ orbitals, and the ligands can move in closer, thereby reducing the ionic radius (see Figure 2.8). In molecular orbital theory, the $t_{2g}$ electrons are non-bonding and the $e_g^*$ electrons are antibonding. The presence of electrons in $e_g^*$ will increase the radius. For $t_{2g}^5 e_g^{*2}$, there are less electrons in the $e_g^*$ level than if they were equally distributed as in the reference state (Section 9, p. 47). Hence the radius will be smaller than that expected for an equal distribution of electrons over $t_{2g}$ and $e_g^*$.

### Q3 (Learning Outcomes 2 and 3)

In $CoCl_2$, $Co^{2+}$ will be in an octahedral environment with configuration $t_{2g}^5 e_g^2$. It therefore has a CFSE of $\frac{4}{5}\Delta_o$ compared with a $d^7$ spherical ion, and will gain this much energy in the crystalline state. Hence, the lattice energy is lower than that predicted for a spherical ion (see Figure 2.9).

### Q4 (Learning Outcomes 2 and 5)

A strong-field octahedral $Co^{2+}$ ion will have a configuration $t_{2g}^6 e_g^1$. With only one electron in the $e_g$ level, this will be a degenerate state; according to the Jahn–Teller theorem, it will distort. In $\mathbf{D}_{4h}$, the $e_g$ will split into $a_{1g}$ and $b_{1g}$, and the electron will occupy the lower level. A distortion in which either two cobalt–ligand distances are shorter, or one in which two cobalt–ligand distances are longer than the other four would remove the degeneracy; it is not possible to predict which will occur in practice.

## Q5 (Learning Outcomes 2 and 6)

Square-planar complexes of cobalt(II) will have one unpaired electron, whereas tetrahedral complexes have the configuration $e^4t_2^3$, with three unpaired electrons. The magnetic moments depend on the number of unpaired electrons, and using the spin-only formula (Equation 6.9) we would expect square-planar complexes to have moments close to $1.73\mu_B$, and tetrahedral complexes to have moments close to $3.87\mu_B$. The observed values are higher than these, but nevertheless indicate that the complexes have the predicted number of unpaired electrons.

## Q6 (Learning Outcome 2)

Figure Q.6 shows the orbitals. The appropriate empty $\pi$-bonding orbitals on CO are the $2\pi^*$. The same combination of these orbitals overlaps with the Co $3d_{xy}$ orbital to form $t_{2g}$ and $t_{2g}^*$, but lobes of the same sign overlap in the case of $t_{2g}$, and of opposite sign in the case of $t_{2g}^*$.

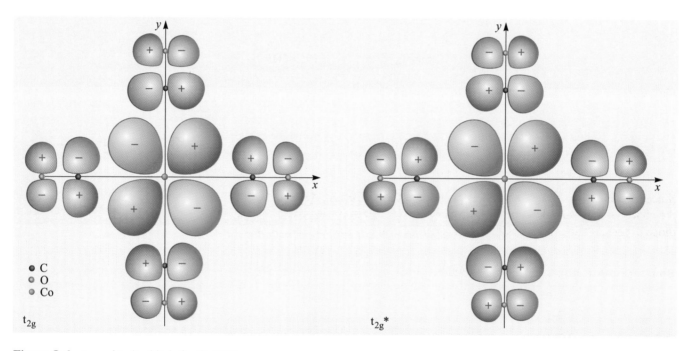

**Figure Q.6**  $t_{2g}$ and $t_{2g}^*$ orbitals for $Co(CO)_6$.

## Q7 (Learning Outcomes 2 and 9)

The orbital energy-level diagram for $[Co(CN)_6]^{3-}$ will resemble that for $[Fe(CN)_6]^{3-}$ (Figure 9.13), but with $Co^{3+}$ ($d^6$) replacing $Fe^{3+}$ ($d^5$). Figure Q.7 shows the partial orbital energy-level diagram.

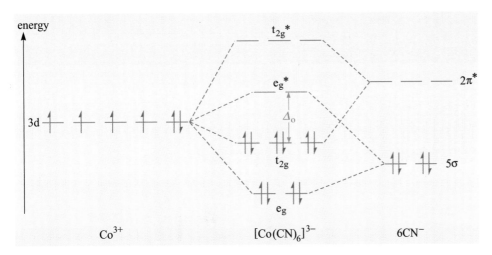

**Figure Q.7**  Partial energy-level diagram for $[Co(CN)_6]^{3-}$.

## Q8 (Learning Outcomes 4, 9 and 11)

For $Cl^-$, $\Delta_0 \approx 5\,600\,cm^{-1} = (45\,000 - 39\,400)\,cm^{-1}$.

For $Br^-$, $\Delta_0 \approx 4\,700\,cm^{-1} = (35\,300 - 30\,600)\,cm^{-1}$.

Since the ionisation energy of Cl 3p is greater than that of Br 4p, we would expect the charge-transfer bands of the $Co^{2+}$ ion in LiCl to be at higher wavenumber than those of the $Co^{2+}$ ion in LiBr. $Cl^-$ is a stronger-field ligand than $Br^-$, so we would expect $\Delta_0(Cl) > \Delta_0(Br)$.

Both expectations are fulfilled by the data.

## Q9 (Learning Outcomes 2, 6 and 9)

$NH_3$ and $Cl^-$ are both ligands with filled $\sigma$- and $\pi$-bonding orbitals. However, this complex has two *trans* ligands of stronger field than the other four. This means that $d_{z^2}$ is bound more strongly than $d_{x^2-y^2}$ and $d_{xz}$ and $d_{yz}$ are bound more strongly than $d_{xy}$. Hence the $a_{1g}$ level will be lower than the $b_{1g}$, and the $e_g$ will be lower than the $b_{2g}$. The $\sigma$-bonded levels ($a_{1g}$ and $b_{1g}$) are still lower in energy than the $\pi$-bonded levels. The bonding orbitals will thus be in the order $a_{1g}$, $b_{1g}$, $e_g$, $b_{2g}$. As in Figure 10.4, the ligand-field orbitals are the antibonding orbitals, and these will lie in the order $b_{2g}{}^*$, $e_g{}^*$, $b_{1g}{}^*$, $a_{1g}{}^*$. The electronic configuration of the complex will therefore be $b_{2g}{}^{*2} e_g{}^{*4}$.

# ACKNOWLEDGMENTS

Grateful acknowledgement is made to the following sources for permission to reproduce material in this book:

*Figure 3.5a*: collection of Professor H. Bank, Idar-Oberstein, Germany, courtesy of the Mineralogical Museum, Würzburg; *Figure 3.5b*: courtesy of the Mineralogical Museum, University of Würzburg; *Figure 5.6*: courtesy of Yvonne Ashmore; *Figure 14.5*: reproduced from the Protein Data Bank, PDB ID: 5AZU, published in H. Nar *et al.* (1991) 'Crystal structure analysis of oxidized *Pseudomonas aeruginosa* azurin at pH 5.5 and 9.0. A pH-induced conformational transition involves a peptide bond flip', *Journal of Molecular Biology*, 221, pp. 765–72.

Every effort has been made to contact copyright holders. If any have been inadvertently overlooked, the publishers will be pleased to make the necessary arrangements at the first opportunity.

# INDEX

Principal references and definitions are indicated by bold type. Numbers and Greek letters are listed as if they were spelt out in full, as they are commonly spoken in English; for example, 'three d' appears as if it were '3d', and 'π' as 'pi' .

# P

# Q

# R

# S